THE Metabolism OF Settlement Coexistences

ORO Editions
Publishers of Architecture, Art, and Design
Gordon Goff: Publisher

www.oroeditions.com
info@oroeditions.com

Published by ORO Editions

Graphic Design: Pablo Mandel, Micaela Carraro / CircularStudio
ORO Managing Editor: Kirby Anderson

Typeset in Inter and Darkmode Mono

10 9 8 7 6 5 4 3 2 1 First Edition

Library of Congress data available upon request. World Rights: Available

ISBN: 978-1-961856-17-2

Color Separations and Printing: ORO Editions, Inc.
Printed in China.

International Distribution: www.oroeditions.com/distribution

ORO Editions makes a continuous effort to minimize the overall carbon footprint of its publications. As part of this goal, ORO Editions, in association with Global ReLeaf, arranges to plant trees to replace those used in the manufacturing of the paper produced for its books. Global ReLeaf is an international campaign run by American Forests, one of the world's oldest nonprofit conservation organizations. Global ReLeaf is American Forests' education and action program that helps individuals, organizations, agencies, and corporations improve the local and global environment by planting and caring for trees.

THE Metabolism OF Settlement Coexistences

PETER G. ROWE, CLAIRE DOUSSARD and **CARLOS ARNAIZ**

with Yoeun Chung, Pauline Detavernier, Yun Fu, Rolando Girodengo, Boya Guo, Priyanka Kar, and Elyjana Roach

ORO
EDITIONS

01 Introduction

This book is about a method for describing and assessing the metabolic contributions of various kinds of settlement coexistences, or basic manners in which humankind assembles itself into urban and quasi-urban constructed circumstances. In what follows there is a brief account of the book's scope, relation to Anthropogenic Era policy-making initiatives, and the organization of the remainder of the book itself.

A. Scope.

With the onset of the Anthropocene Era, concern for the metabolism of various kinds of urban settlement has risen appreciably. As will be discussed, it has expanded well beyond earlier physiological accounts of humankind into the multitude of activities and settings of modern life. Of particular concern in the study of architecture and urban design is the metabolic contributions of flows of stocks that go into the building and operation of settlement collectives of one kind or another. What is involved here and dealt with in this volume is an approach to rewriting, as it were, urban settlement patterns in terms of their water, energy, and material flows in order to use the resulting depictions as bases for proposing technological and design proposals for ameliorating and diminishing adverse effects and impacts, as well as related outcomes such as embodied energy and carbon concentrations so deleterious to climate change, as well as proliferation of other hyperobjects. Accordingly, hyperobjects are those that are so massively distributed in time and space as to transcend spatiotemporal specificity (Morton, 2010; and Morton, 2013). They include such processes as 'global

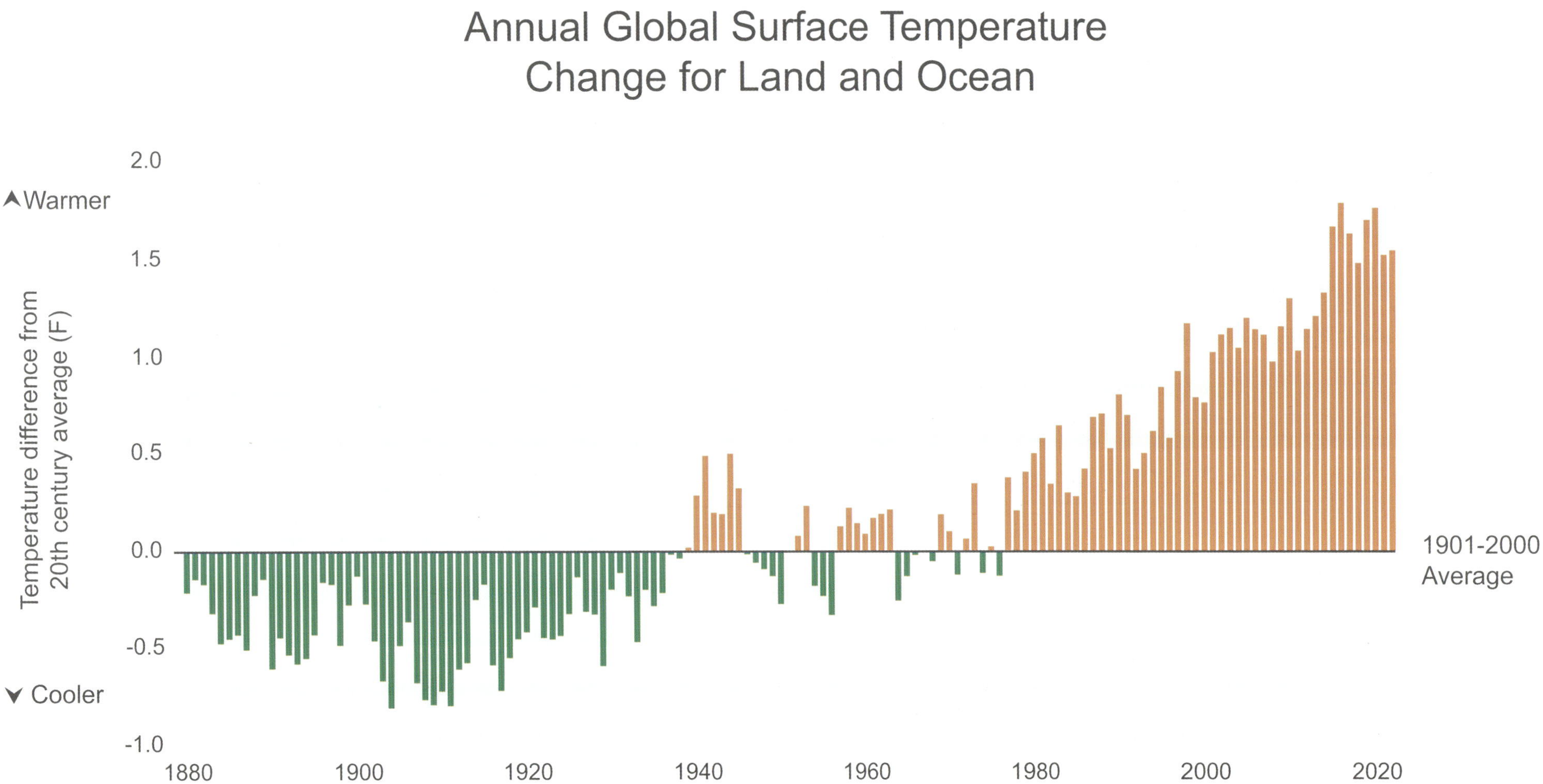

FIG. 1: Global Average Surface Temperature.

warming' and object accumulations like 'Styrofoam' and others like 'radioactive plutonium.' Thinking along these lines has also given rise to 'object-oriented-ontology' with its rejection of the anthropomorphism that privileges humans as subjects against others as objects and is defined by the likes of Gregory Harman, the American speculative philosopher (Harman, 2011).

B. Policy-Making Initiatives.

International policy-making initiatives, largely within the purview of the United Nations, began as early as 1972 with the Stockholm 'United Nations Conference on the Human Environment' and on the heels of 'Earth Day' organized by Denis Hayes in 1970 and the birth of the modern environmental movement, particularly in the United States (Waxman, 2019). However, it wasn't until some 20 years later that more concerted action began to take place under the United Nations at the Rio Earth Summit of 1992. Shortly before that, groundwork was laid in 1987 with the so-called 'Brundtland Report,' published in recognition of Gro Harlem Brundtland, the former Norwegian Prime Minister and chairperson of the World Commission on Environment and Development (World Commission on Environment and Development, 1987). The Rio Summit was followed in 1997 by high-level meetings at the UN General Assembly in New York and, later, in 2002 by the Johannesburg 'World Summit on Sustainable Development' with follow-up reviews, again in New York in 2005, 2008, and 2010. A second Rio Conference was held in 2012 on sustainable development, followed again by follow-ups in New York in 2015 establishing the 2030 sustainability targets alongside the Paris Agreement also in 2015. Finally, in 2022 Stockholm again hosted a conference, also addressed to the 2030 targets. Over the years the matter of 'Climate Change' emerged and influenced policymaking in particular with increasing strength. Scientific data and various temperature benchmarks began to be established in conjunction with this world-wide threat, with the U.S. Global Change Research Program estimating longer-term rises as much as 2.7°C to 5.6°C above 1901 or pre-industrial averages, and as shown in Figure 2 (Wuebbles, Fahey, Hibbard, et al., 2017). While the topic of climate change remains controversial, politically anyway, recent comments by Professor Jim Skea,

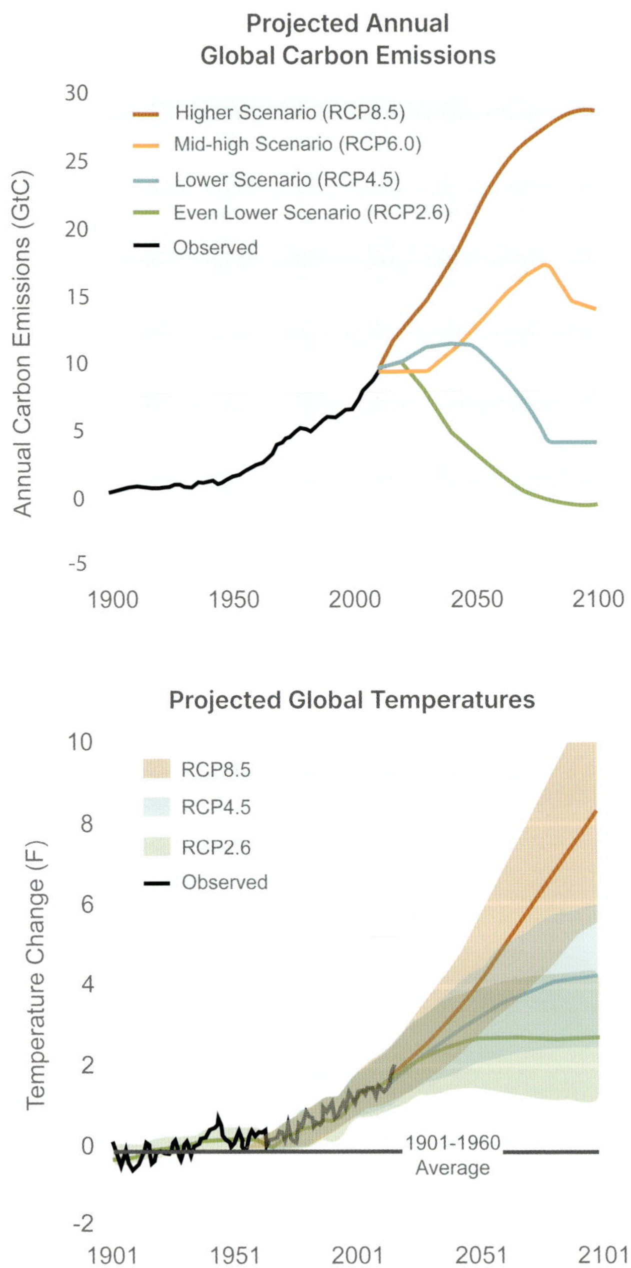

FIG.2: Projected Global Carbon Emissions and Temperatures.

the Scot elected to head the 'UN Intergovernmental Panel on Climate Change,' have also been directed towards a certain sense of realism regarding the potential existential threat of climate change, referring instead to the specter of a soon-to-be-reached 1.5°C rise above pre-industrial levels as more of a warning than an outright threat (Sommerville, 2023). Nevertheless, evidence of the persistence of climate change, largely due to carbon emissions into the atmosphere, has led to rises of 1°C since the late 18th century, to oceans getting warmer in the top 100 meters of water, and to ice sheets melting and sea levels rising at around 20 centimeters over the past century or so. Moreover, consumption of natural resources more generally has accelerated and contributed to dire accounts of the Anthropogenic Era, alongside of the need for higher awareness of the rates of consumption and the metabolic contributions of various forms of human settlement (Turvey and Crees, 2019).

C. Organization of the Book.

What follows is a brief historical account of metabolic depictions, beginning in the 16th through 18th centuries, with the work of Santorio and Lavoisier, and moving through the principles of mass and energy conservation into the Anthropocene Era and concepts of the metabolism of cities. This is followed by discussion of stock-flow modeling of water, energy, and materials in conjunction with various kinds of settlement coexistence from extraction from natural domains like the geosphere, biosphere, and hydrosphere; through construction and operation of settlement developments; and on into end-of-life scenarios involving refurbishing, reuse, and mining from demolition activities. Case studies follow, each associated with a particular form of settlement coexistence. Here, compact and central city areas are represented by the special zone (ZAC) of Clichy-Batignolles in Paris, France. The dispersed city and peripheral urban development are represented by the Westwood community on the outskirts of metropolitan Boston in the United States. Organic and informal settlement is taken up for the Colonia La Campana in Monterrey, Mexico, and desakota development by the township-village enterprise of Chenchun in southeast China. Each of these communities are cases of roughly the same population size of around 7,000 inhabitants, also allowing for some comparative analysis. These case studies are augmented by the study of a specific commercial district on the Champs-Élysées in Paris and typical housing types among the four domains of settlement coexistence, as well as a proposed project in Victorias in the Philippines. Throughout, an emphasis is placed on methodological aspects of metabolic stock-flow modeling and comparative analyses if results were appropriate.

02

A Brief Historical Account of Metabolic Depictions of Humankind

The perception of metabolic phenomena, particularly with regard to humankind, more or less began with the research of Santorio Santorio (1561–1636), a professor at the University of Padua in Italy and generally considered the founder of human metabolism (Santorio, 1737). He had taken it on himself to measure his bodily physical processes by weighing his food intake, his body weight, and the weight of the excreta he passed on a daily basis. His resulting input-output balance over a long period of time revealed that about half of the input was not found in the increase in body stock nor output via excreta. He then went on to conclude that this discrepancy was due to some 'invisible perspiration.' Sometime later, Antoine Laurent Lavoisier (1743–1794), the French father of modern chemistry, also made laboratory experiments, weighing all educts and products of controlled chemical reactions and leading to the falsification of the then-prevalent 'phlogiston theory,' which posited that this phlogiston substance was a component that combusted and explained the loss or gain in weight observed by the likes of Santorio, and where the term derived from new Latin from Greek meaning to set alight or burn. This then led Lavoisier to unravel the mystery or missing link in human metabolism in which the mass and chemical composition of respiratory gas could be measured and identified as carbon dioxide. More generally and more importantly, it led to his 'Law of Mass Conservation,' by which, in a closed system, the mass of the products in a chemical reaction equals the mass of the reactants (Lavoisier, 1789). It led further to 'stoichiometry,' or the relationship between quantities of reactants and products before, during, and following chemical reactions (Lavoisier and Laplace, 1780).

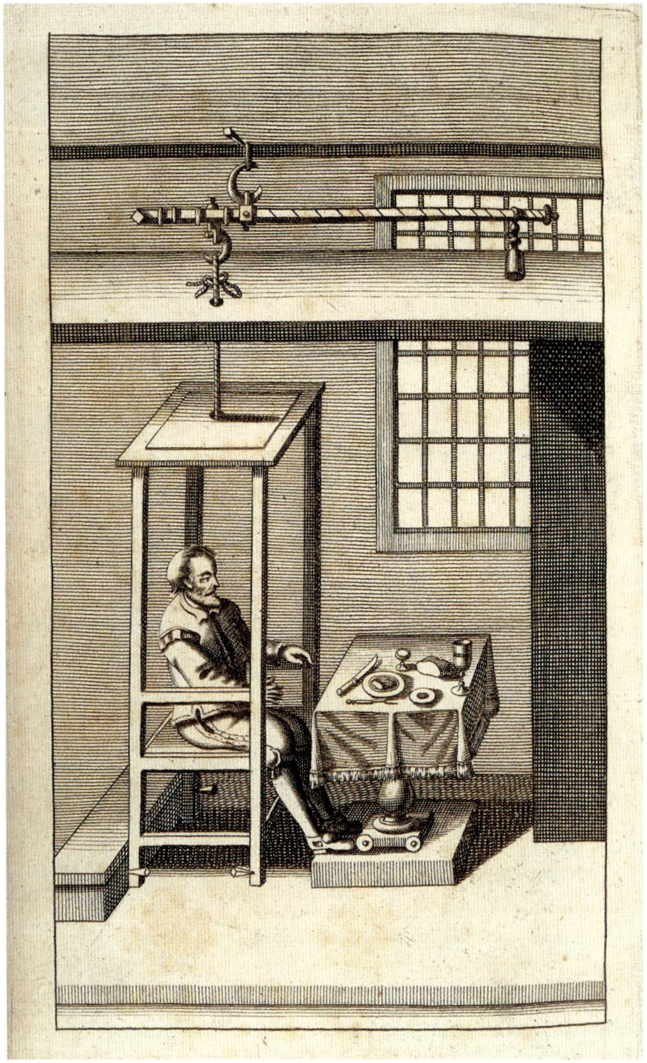

FIG. 1: Santorio's Weighing Apparatus.

Also, during the 18th century, the 'Conservation of Energy Principle' was first enunciated by Emilie, Marquise du Châtelet (1706–1749), the French philosopher and mathematician who both proposed and tested the principle, apparently inspired by the theories of Gottfried Leibnitz (1646–1716), the German Polymath (Marquise du Châtelet, 1744). Shortly before she died, she published a French translation with commentary of Isaac Newton's *Philosophiae Naturalis Principia Mathematica*, including her derivation of the conservation of energy from Newton's principles of mechanics (Lawson, Hostetler, and Edwards, 2008). Like the Law of Mass Conservation, the Conservation of Energy Principle states that the total energy of a closed system remains constant, with energy being only transferred or transformed and not created or destroyed. One of the bases for understanding this principle is Noether's Theorem developed in 1915 and named for Emmy Noether (1882–1935), the important German mathematician (Neuenschwander, 2011). Beyond its continued evolution in physics, this and other related principles became important parts of material balance studies associated later with space travel, leading also towards ideas about controlled ecological life-support systems. Inherent here is also the distinction between 'closed,' 'open,' and other kinds of systems. Moreover, there is some variation depending upon the intellectual discipline involved. In thermodynamics, that branch of physics which explains energy transfers between objects and surroundings, there are three kinds. The are open, closed, and isolated systems, where 'open systems' allow the exchange of both energy and matter with surroundings; 'closed systems' that do not allow the transfer of matter but do permit energy transfers; and 'isolated systems,' as the term suggests, that do not allow either matter or energy transfers to occur (Hewitt, Suchcock, and Hewitt, 2009). However, a closed system in classical mechanics is equivalent to an isolated system in thermodynamics, whereas a closed system in quantum mechanics is one that does not allow any information exchange with another system. Continuing on, in sociology open systems are ones that allow mixing between social classes, for instance, defined by measures of social stratification such as wealth, ancestry, and so on (Parsons, 1974). In this account here, the engineering definition of a closed system is adopted as one in which every input is known, and every resultant is also known or able to be known. Indeed, it is one of the strict aspects of stock-flow modeling. Simply put, what goes in is what comes out in mass, though in different form (Bertalanffy, 1968).

Nowadays, however, unlike Santorio's metabolic experiments that focused on a crude physiologic perspective of human life, modern humankind relies on many more materials for their activities, as they appear to have entered the era of the Anthropocene (Baccini and Brunner, 2012). This era is still somewhat controversial and not fully accepted. It has been proposed as a geologic era dating from the commencement of significant human impact on the Earth's geology and ecosystems, probably around the end of World War II and the beginning of the proliferation of atomic energy, including but not limited to Climate Change. Though dating back to use by Soviet scientists in the 1960s to refer to the Quaternary as the most recent geologic period, the term 'Anthropocene' was widely popularized by Paul Crutzen (1933–2021), the Dutch Nobel Laureate, meteorologist, and atmospheric chemist. In this Anthropocene concept, he and others argued for a more holistic view of the role of humans in interacting and shaping natural systems. In general, this did not follow strictly according to a narrative occurring fully in a declining direction. Instead, they argued for a broader perspective that also admitted achievements in human-natural system interactions (Steffen, Crutzen, and McNeill, 2007).

The more specific idea of the metabolism of the city was mooted by Abel Wolman as early as 1965 (Wolman, 1965). One of the advantages of this metabolic approach is that it comprehensively links material characteristics of the natural world and domains like the geosphere, biosphere, hydrosphere, and so on, with otherwise conveniently defined land-use types and activities of the 'technosphere,' together with the residuals of spatial settlement production in the form of waste, recycling potential, and management. It also does so in a manner that directly reflects temporal variation in inevitable processes like deterioration and obsolescence and is almost entirely object-oriented in its ontology (Harman, 2018). Several factors have influenced this phenomenon of anthropogenic metabolism in settlement. The first is increased levels of urbanization with now over 50 percent of the world's total population living in urban circumstances of one kind or another. While impacts might have

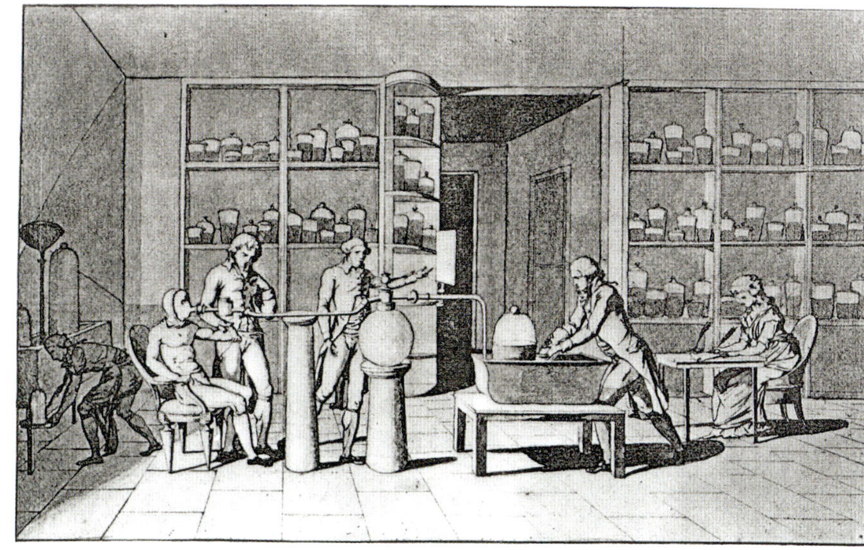

FIG. 2: Lavoisier Conducting an Experiment on Respiration.

diminished due to more efficient networks, the overall higher density of living has exerted higher pressure on environments because of all the nutrients that now must be dispersed into water, soil, and air and especially carbon contributions to the atmosphere and subsequent effects on other phenomena like climate change. Second, with regard to the dynamics of flow rates of material use from year to year, they have risen appreciably for products like concrete, paper, and plastic, as well as for manufactured materials like aluminum and those with zinc and lead bases (Baccini and Brunner, 2012). Third, as inputs to settled areas have exceeded exports, stock formation has increased. Moreover, these stocks are often important reservoirs of substances requiring attention to recycling and urban mining. In addition, they are also often sources of hazardous substances and pollutants. Fourthly, over time, consumption emissions have surpassed production emissions with the upshot that prevention of consumer-related emissions is more difficult to prevent than industrial pollution. Moreover, anthropogenic material flows appear to have exceeded geogenic flows, where the latter derive primarily as a result of natural seepage, soil-wind erosion, and presences of trace elements, such as might be found in geophysical fluid dynamics (Cushman-Roisin and Beckers, 2011; and Baccini and Brunner, 2012). At root, urban metabolism is bound up with the depiction, modeling, and analysis of the flows of materials including water and energy associated with human settlement from its onset in 1965 or so to around 1980. As an area of knowledge and expertise it went through an initial period, followed by a period of steady growth and development (Zhang, Yang, and Yu, 2015). During the late 1970s, for instance, the American 'National Aeronautical and Space Administration' (NASA) published a working document detailing how a system of space colonization could be constructed (Johnson and Holbrow, 1977). This work included stock-flow designs and diagrams concerned with material supplies like water. It also grew out of the foundation of NASA in 1958 and its activities at the Langley Research Center in the Washington, DC, area, beginning in 1961 (Wente, 1990; and Dick, 2008).

Conspicuous during the initial phases of work on urban metabolism up to at least the end of the 1960s, was that performed by H. T. Odum with regard to energy flows. Howard T. Odum (1924–2002) was an American systems ecologist who used an analog of electrical energy networks to model the energy flow pathways of ecosystems. Early in the 1950s he developed an analog of Ohm's Law which aimed to represent energy flows through ecosystems (Kangas, 2004). He also co-authored the *Fundamentals of Ecology* with his older brother Eugene Odum, that paved the way for the popularity of ecology and ecosystem thinking (Odum and Odum, 1953). Later on, material flow analysis (MFA) and substance flow analysis (SFA) were introduced by researchers such as Peter Baccini and Paul Brunner, as well as other work on ecosystem services (Elliot, Almenar, and Niza, 2019). Baccini (1939–) is Professor Emeritus at the Federal Institute of Technology, Zurich ETH where he served as Professor of Resources and Waste Management. This approach bore certain similarities to Wolman's initial proposition from 1965 of a hypothetical American city of one million inhabitants. Baccini's results and commentary were first published in 1991 by Springer Verlag under the same title of *Metabolism of the Anthropocene* (1991). The distinction between material flows and substance flows is that the former describes the anabolic processes of urban settlement in relatively practical terms incorporating building materials, whereas substance flows involved a rewriting, so to speak, of these material flows as chemical substances. The authors after all were chemists among other things (Baccini and Brunner, 2012). They and others also emphasized the relationship of metabolic analysis and urban planning and design (Oswald, Baccini, and Michaeli, 2002; and Kennedy, Pincetl, and Bunje, 2011). In addition, the spatial aspect of urban metabolism was taken up by Bahers and others (Bahers, Athanassiadis, Perotti, et al., 2022. Also closely related to the study of urban metabolism is work with regard to 'circular economies' essentially as models of production and consumption which involve sharing, reusing, refurbishing, and recycling of existing materials and products for as long as possible. Although this concept cannot be ascribed to a particular date or individual it resides in the work done in several quarters. Herbert Girardet (1943–), a German-British commentator, was known to have used the term 'circular' as distinct from 'linear,' aligning with proponents of 'industrial ecology' where they postulated the use of outputs from one process as the inputs of another, as seen at the Kalundborg Industrial Park in Denmark (Frosch and Gallopoilos, 1989). Also among the first users of the term was the economist Alan Kneese when commenting on the economic value of natural resources (Kneese, 1988). In the context of this volume there are at least four lines of circularity involved in constructed environments. Two are involved with maintaining building assets in place. They are end-user maintenance and repair as well as fuller-scale refurbishment. Beyond this is repurposing or reuse of an asset for another purpose, as well as demolition and mining of materials to recycle them in new construction. Inclusion of these cycles refers to 'cradle-to-use' depictions of urban metabolism, as well as 'cradle-to-cradle' depictions (McDonough, 2002).

Also, during the 1960s the discipline of 'ecology' gained substantial public attention, primarily because of widespread concern for the state of the environment. However, ecology is not a new field, with the term first being coined by Ernst Haeckel in 1869. Etymologically, the term derives from the Greek *oikos* meaning 'household' and *logos* meaning 'the study of,' or more fully, 'the study of nature's household.' Haeckel (1854–1919) was a German zoologist, naturalist, and biologist who made major contributions to the study of the form or morphology of organisms of various kinds. Much later and after his death he became tangentially associated with the ugly Nazi side of eugenics, though no hard evidence suggested he was racist (Haeckel, 1866, and Desmond, 1989). Ecology and ecosystem thinking essentially seeks to define the relationships between 'biotic' and 'abiotic' factors in an environmental system. In this regard, ecology is the study of such ecosystems, and abiotic factors are regarded as those that are physical and chemical rather than biological, or, in other words, non-living factors such as water, temperature, acidity, and so on (Rickerts, 2005). Pioneers in this kind of relational thinking included Frederick Clements (1874–1945), the eminent American plant ecologist who produced a monumental study of plants and plant succession as early as 1916, though with considerable prior methodological work as well (Pound and Clements, 1898; Hagen, 1992; and Real and Brown, 2012). Clements' ecological theory rested on two propositions. They were the concept of ecological succession, and treatment of the plant community niche as a 'complex organism' undergoing a cycle of evolution. Others, such as Gleason and Tansley, disputed some of this and revised Clements' concepts (Real and Brown,

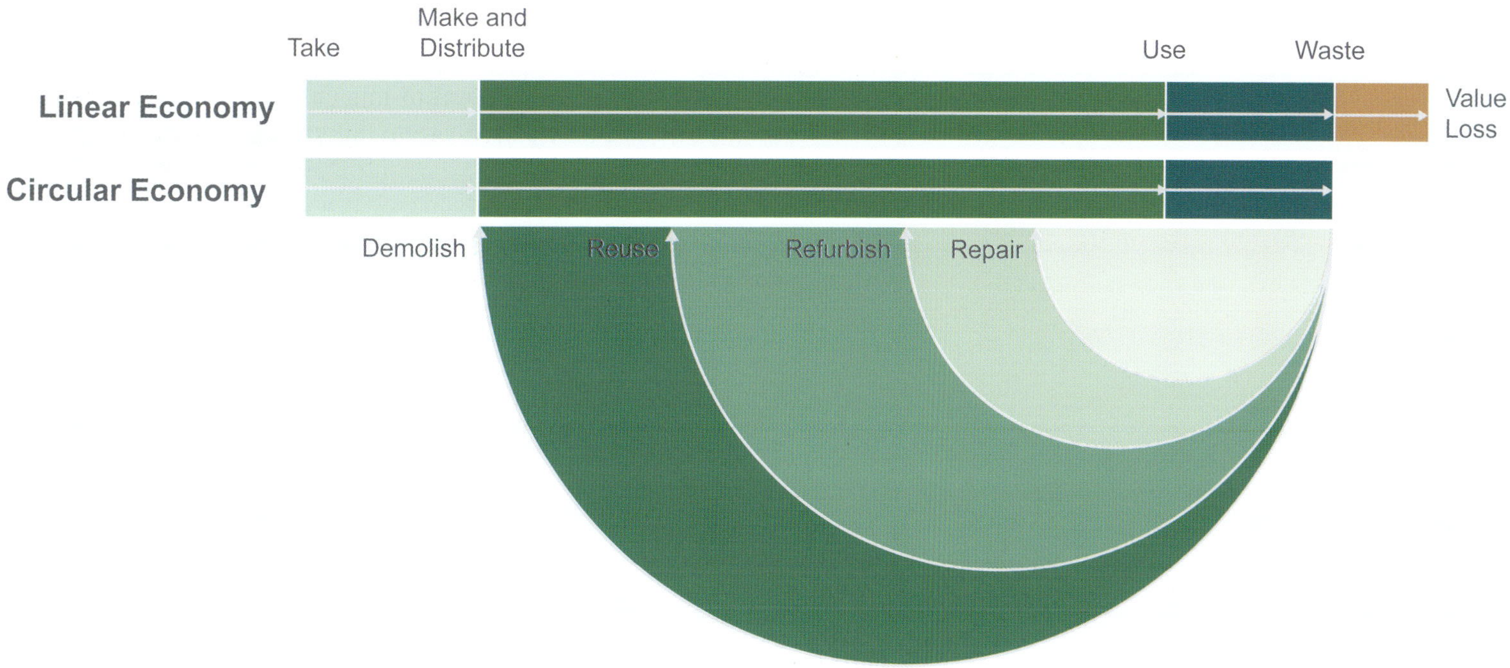

2012). As noted earlier, ecosystem services are ecologically mediated functional processes essential to sustaining healthy human societies and public health. These include processes like water provision and filtration, agriculture and fisheries, and the removal of greenhouse gases from the atmosphere. Failure to allow these processes to operate naturally or ecologically can run the risk of alienating ecosystem processes from serving public health (Elliot, Almenar, and Niza, 2019). Finally, at least within the design disciplines, 'landscape ecology' has emerged as a strong set of principles. The term was first coined in 1939 by the German geographer Carl Troll (1889–1975) as *landschaftsökologie* (Gade, 1996). Also critical to landscape ecology theory were works by Edward Wilson, Richard MacArthur, and Richard Forman (MacArthur and Wilson, 1967; and Forman, 1995). During this later period the use of digital data and geographic information systems also played a significant role in advancing ecological thinking spatially (Steinitz, 2012).

FIG. 3: Circular vs. Linear Relationships in Urban Settlements.

03

Stock-Flow Modeling and Sankey Diagrams

What follows is an account of the stock-flow modeling process pursued in these research and development exercises. With reference to the preceding section of this volume, it is a stock-flow model that proceeds from 'cradle-to-grave,' via intermediate material extraction, construction, and land-use settlement on to 'end-of-life' activities. In this context the initial starting point is a more or less fully developed settlement. In other words, the stock-flow relates to a completed project and deals, therefore, with its initial material composition and operating characteristics. Further, this account begins with a definition of domains of settlement coexistence dealt with by the model and case study exercises, followed by a similar background for the natural domains involved in the modeling from which materials are extracted to form stocks that then flow through the construction, operation, and end-of-life of settlements. The account of this last phase, broadly under 'waste management,' dwells primarily on the matter of 'obsolescence' and ascription of an annual rate of obsolescence that is so fundamental to the stock-flow sequence, and also taken up again in the final Chapter 6 of this volume. It is, after all, an expression of material weight or volume per year. Commentary is also offered about data sources, typically of three kinds, including direct observation, data obtaining to immediate site surroundings, and proxies drawn from broader surrounding jurisdictions. Representation of results primarily takes the form of Sankey diagrams that graphically depict flows of materials from their natural sources through construction and operational phases to end-of-life management. The case studies that follow this account illustrate particular applications of stock-flow modeling.

A. Domains of Settlement Coexistence.

Urban settlement is and has been variously defined. For some it follows core-peripheral patterns of relative building density, like 'central business districts,' 'inner suburbs,' and peripheral areas. For others it is seen to be evolutionary, involving stages of development from villages to towns and into cities. For still others the core-periphery arrangement is subdivided into categories like 'downtown loop,' 'factory zone,' 'zone of transition,' commuter zone,' and so on (Storper, 1997; Hanlon, Short, and Vicino, 2010; and Brenner, 2019). More fundamentally though, people can be seen to exist in sustainable though dynamic states by sharing built landscapes within tolerable risks and, thus, sharing common coexistences (Klouche, 2010). Among such coexistences at least four stand out. First, there is the 'compact city' of relatively dense urban areas coinciding with most if not all accepted definitions of being in a city, such as in Manhattan, Hausmann's Paris, or Cerda's Barcelona. Second, there is the 'dispersed city,' often commonly known as the suburbs. In the U.S., for instance, there are 'automobile suburbs,' particularly adjacent to the Interstate Highway system, their peripheral areas often encompassing once rural towns. Third, there is the 'organic city' coinciding with informal settlement and coexistence by migratory and sometimes poorer populations at points in time, such as the *barrios independenza* in Monterrey, Mexico, like Colonia La Campana. Fourth, there are 'desakota areas,' from the Indonesian *desa*, meaning village, and *kota*, meaning town, which usually arise adjacent to established urban areas but within which inhabitants engage in both rural and non-rural activities by way of livelihood. In many ways they are also interpretations of urban-rural continua in Indonesia or township-village enterprises in China. In summary, the modeling efforts described here deal with all four of these settlement coexistences of the following kinds:

1. Compact City – Pre-modern and modern developments in central city areas as well as contemporary eco and special community redevelopments in those areas;
2. Dispersed City – Automobile suburbs and peripheral urban developments within metropolitan areas;

3. Organic City – Legally informal areas, often self-built areas and makeshift shanty towns often in peripheral areas of established cities; and
4. Desakota Areas – Combining both urban and agricultural activities, sometimes with industry and often associated with more compact settlement areas.

Figure 1 depicts differences among these and other spatial urban configurations.

B. Natural Domains.

On the other side of a 'settlement-nature' arrangement are the natural domains from which materials introduced into settlements are derived. There are five such domains in all, each of which is complex and interacts with other domains. For instance, the biosphere of living beings is supported by the geosphere of the interior and exterior of the earth, as well as the hydrosphere of water. The five main systems are the geosphere; the biosphere; the hydrosphere; the atmosphere, or envelope of gas that keeps the planet warm and provides oxygen for respiration; and the cryosphere, comprised of huge amounts of ice, primarily at the two poles. Many building materials like sand, limestone, and even oil products are sourced from the geosphere, whereas timber and organic materials are sourced from the biosphere.

C. Model Framing.

The basic framing of the stock-flow model can be represented on a simple linear relational diagram with some circularity, as shown in Figure 2. As indicated, it begins on the far left-hand side with the 'resource domains' of, for instance, the geosphere, biosphere, and hydrosphere and then moves through a sequence of major processes before emerging into the 'end-of-life phase' of resource accounting. These major phases are at least sixfold and begin with the 'extraction phase' from the natural domains expressed in mass or volume terms per year and as relative proportions of overall resource consumption. This is followed by the 'material production phase,' resulting in building materials and/or energy and water consumption. These materials, in

FIG.1: Spatial Depiction of Different Domains of Urban Settlement.

turn, are then made into constructed components during the 'constructed environments phase' such as buildings, elements of infrastructure, and varieties of non-built spaces, which are then grouped in the 'land-use phase.' Generally, among these uses are residential, commercial, and public uses; industrial and agricultural uses; as well as infrastructure and managed alongside natural green uses. Then the stock-flow process enters something of a steady state referred to here as the 'stock phase,' consisting of the uses that remain longer through time and before entry into the 'end-of-life phase' in which there are at least four consistent and often simultaneous outcomes, ranging through 'repair,' 'refurbishing,' and 'reuse,' which extend the life of constructed elements, and finally 'demolition.' Demolition, in turn, is further subdivided into a the 'waste management phase' with at least two outcomes. They are 'waste handling' including recycling, sorting, and composting, followed by 'waste disposal' including mining, landfilling, and incineration. Of these, mining is becoming more crucial as municipalities and other responsible institutions are attempting to actively increase levels of reusable material resources (Diehl and Smith, 2003). As will be discussed later on, the flow rates in units per time and largely embracing obsolescence rates are applied throughout.

D. Stocks.

Strictly speaking, material stocks are holdings or commodities that result from the production of core or primary materials used by other industries or modes of production. They are largely involved in processing or discovering raw materials like oil, stone, or chemicals. In fact, the materials sector is one of 11 major sectors of the stock market. Anthropogenic stocks, of the kind accounted for in this volume, can be reused and are essential components of other made-materials like plastics, fertilizer, paper, concrete, and metal products. For the most part, anthropogenic stocks are deployed in settlement construction and to support lifestyle activities. As such they are described as they emerge from extraction within natural domains to be manufactured or otherwise produced into products like building materials and then combined to produce buildings and elements of infrastructure and then on to form categories of land use. Throughout the modeling exercises

FIG. 2: A Sankey Diagram Incorporating End-of-life Cycle Repair, Refurbishment, Reuse and Demolition.

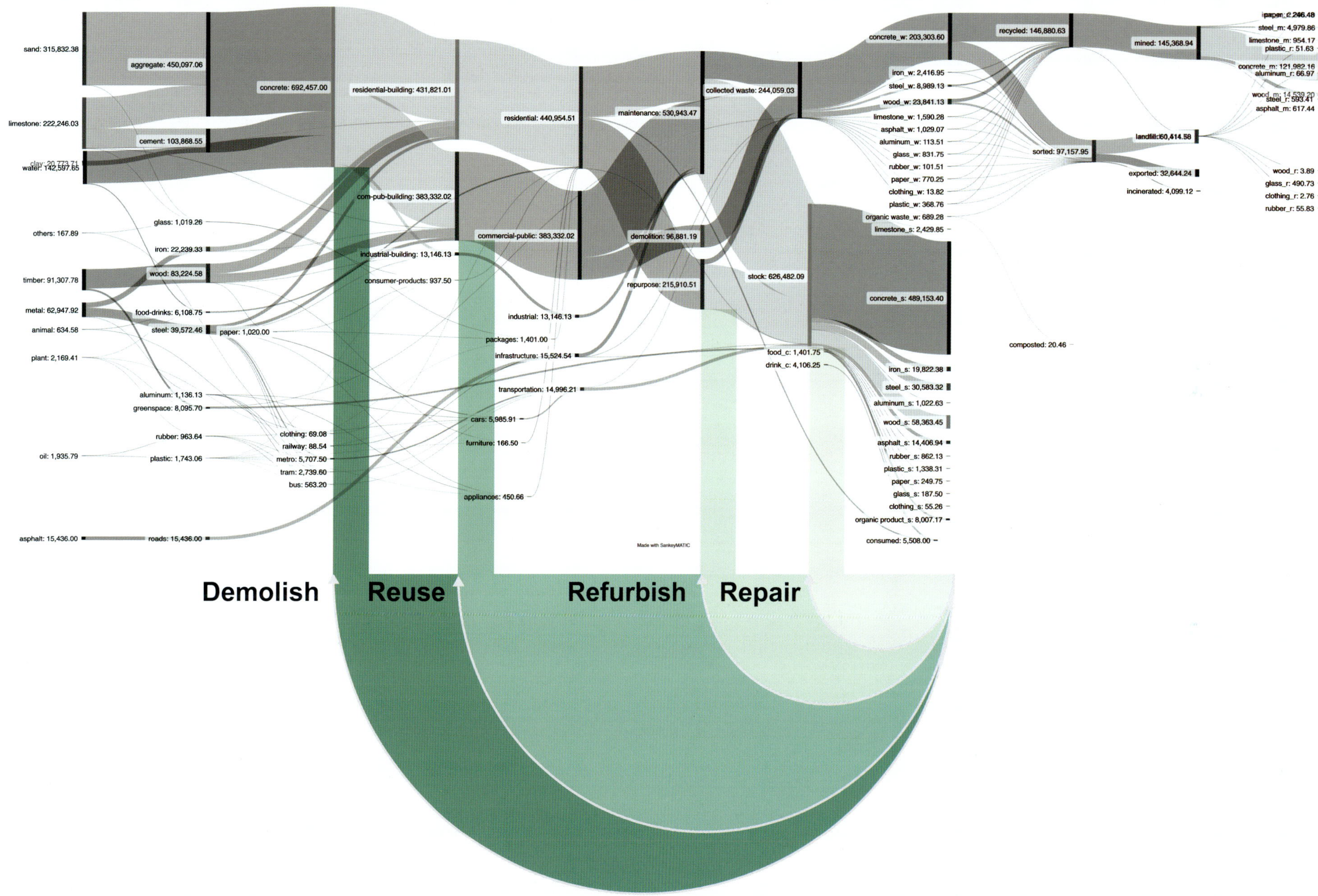
sand: 315,832.38
limestone: 222,246.03
others: 167.89
timber: 91,307.78
metal: 62,947.92
animal: 634.58
plant: 2,169.41
oil: 1,935.79
asphalt: 15,436.00
aggregate: 450,097.06
cement: 103,868.55
glass: 1,019.26
iron: 22,239.33
wood: 83,224.58
food-drinks: 6,108.75
steel: 39,572.46
aluminum: 1,136.13
greenspace: 8,095.70
rubber: 963.64
plastic: 1,743.06
roads: 15,436.00
concrete: 692,457.00
paper: 1,020.00
clothing: 69.08
railway: 88.54
metro: 5,707.50
tram: 2,739.60
bus: 563.20
residential-building: 431,821.01
com-pub-building: 383,332.02
industrial-building: 13,146.13
consumer-products: 937.50
residential: 440,954.51
commercial-public: 383,332.02
industrial: 13,146.13
packages: 1,401.00
infrastructure: 15,524.54
transportation: 14,996.21
cars: 5,985.91
furniture: 166.50
appliances: 450.66
maintenance: 530,943.47
demolition: 96,881.19
repurpose: 215,910.51
collected waste: 244,059.03
stock: 626,482.09
food_c: 1,401.75
drink_c: 4,106.25
concrete_w: 203,303.60
iron_w: 2,416.95
steel_w: 8,989.13
wood_w: 23,841.13
limestone_w: 1,590.28
asphalt_w: 1,029.07
aluminum_w: 113.51
glass_w: 831.75
rubber_w: 101.51
paper_w: 770.25
clothing_w: 13.82
plastic_w: 368.76
organic waste_w: 689.28
limestone_s: 2,429.85
concrete_s: 489,153.40
iron_s: 19,822.38
steel_s: 30,583.32
aluminum_s: 1,022.63
wood_s: 58,363.45
asphalt_s: 14,406.94
rubber_s: 862.13
plastic_s: 1,338.31
paper_s: 249.75
glass_s: 187.50
clothing_s: 55.26
organic product_s: 8,007.17
consumed: 5,508.00
recycled: 146,880.63
sorted: 97,157.95
composted: 20.46
mined: 145,368.94
exported: 32,644.24
incinerated: 4,099.12
steel_m: 4,979.86
limestone_m: 954.17
plastic_r: 51.63
concrete_m: 121,982.16
aluminum_r: 66.97
steel_r: 593.41
asphalt_m: 617.44
wood_r: 3.89
glass_r: 490.73
clothing_r: 2.76
rubber_r: 55.83
Made with SankeyMATIC
Demolish
Reuse
Refurbish
Repair

described here, the volume/weight of each stock component is constant. In other words, inputs are equal to outputs at each stage of the transformational process.

E. Flows.

Material, water, and energy flows in these modeling exercises emerge from sources, as noted, and flow between various steps or points of a supply or value chain. Indeed, the basic framework of a stock-flow model consists of relatively fixed points at which material stocks become transformed into something else and become the inputs into further transformation processes. For instance, oil is extracted from the geosphere, processed into asphalt, and combined with roadbeds to form a 'paved' surface to support vehicular transportation. In these models the principal alignments of these points correspond to extraction; product manufacture; assemblage of manufactured products into, say, constructed buildings like houses, stores, roads, and so on; and then on to full assemblages of land-use like residential, infrastructure, commercial, and industry; as well as on into end-of-life disposal and re-use. This kind of flow can also be described as stage-processing, representing the way that workflows unfold between particular events or categories of production (Sollis-Saether and Gottschalk, 2010).

F. Waste Management.

As noted in the earlier section, waste management now usually involves some degree of circularity. That is to say that certain outputs can become the inputs of other processes involved in settlement activities. Compared to a simple linear flow of stages in a model, there are at least four forms of feedback or circularity involved, as illustrated in Figure 2 and Figure 1 in Chapter 2. The first involves 'end-user maintenance and repairs.' Second, there is fuller fledged 'refurbishment' of the fixed asset, that may also involve extensive resurfacing and remaking, often resulting in interruption of use. Third, there is 'repurposing' or 'reuse' of the asset for another function entirely, such as conversion of commercial or industrial structures into residential apartments. Fourth, there is demolition of the asset and the mining of its remains, so to speak, as a source of material for further new use. Though different, all share in common some reckoning with 'obsolescence' and the process of declining performance due to changes in the physical, functional, economic, social, legal, and political environment in which the asset resides. In this volume useful definitions of obsolescence are presented essentially based on the physical and useful service life of elements of a constructed environment. More generally, the ISO, or International Organization of Standardization, defines obsolescence as the loss of ability to perform satisfactorily due to changes in performance requirements, and is often also considered the beginning of the 'end-of-life' phase of building among a string of stages such as 'development,' usage,' and 'reuse' (Rackwitz, 2000; and ISO, 2015). Also as noted earlier, obsolescence is resisted by repair, refurbishment, reuse, and finally demolition and mining of materials for secondary use (Rowe, 2017). Again, generally according to the ISO, when end-user repair and refurbishment involves more or less one-to-one replacement of an obsolete asset, the annual obsolescence rate is inversely proportional to the lifetime span of the asset, or $w=1/T$, where w is the annual obsolescence rate and T is the estimated lifetime span of the asset. For example, if T was 50 years, then the annual rate of obsolescence would be 2 percent per annum, or 0.02. However, in other formulations the annual obsolescence rate is seen to be modified by an array of up to seven different forms of obsolescence, such as those cited earlier and to be taken up later in Chapter 5 dealing with improved metabolic performance. As noted by others, disparities in these formulations can lead to wide and not very useful variations in obsolescence rates (Diamantidis, Sykora, and Bertacca, 2018). Nevertheless, for the modeling exercises described here a general approach was adopted with some allowable modification of w, the annual obsolescence rate, in which within $w = 1/T (O^i)$, where $O^i=1$ for conditions of high maintenance, 0.9 for normal maintenance, and 0.8 for negligible maintenance. With regard to the other two forms of circularity, reuse or repurposing would involve effectively rerunning the stock-flow model with a different proportion of uses associated with the constructed environment. For demolition, the mining potential of recycled material would be incorporated (Lederer, Laner, and Fellner, 2014; and Cossu and Williams, 2015). Currently in the U.S. some 23.6 percent of municipal solid waste is recycled in the form of minable materials with 17 percent of this in wood, 34 percent in metals, 25 percent in glass, and 22 percent in other secondary materials (Environmental Protection Agency, 2020). In France minable amounts of demolition materials are estimated to be from 60 to 80 percent of collected waste, however cities like Paris are struggling to meet these proportions.

As an example, and with regard to displaying a full representation of the repair, refurbishment, reuse, and demolition phases of end-of-life building activities shown in Figure 2, it begins with estimates of extraction and materialization on the far left-hand side. There is some 73 percent that derives from the geosphere in metric tons per year, 16.3 percent from the hydrosphere, and the remaining 10.7 percent from the biosphere. The estimated lifetime of building assets were 200 years for the Haussmann structures and 100 years for post-1920 buildings, with the exception of those in the commercial-public category at 91 years, based on Atelier Parisian d'Urbanisme (APUR) estimates. Among land uses within the ZAC site 50.3 percent were residential, with 23.8 percent commercial-public and 1.5 percent industrial, alongside 1.8 percent roadways, and green space around 1 percent. Again, these estimates are in material metric tons per year terms. In the transition phase of the model moving towards waste management, maintenance, reuse, and demolition were estimated differently based on the asset type involved. For residential buildings, most recently built, there was no reuse. For commercial-public buildings 2 percent reuse was assumed and 98 percent was under maintenance. Largely this was based on the Ateliers Berthier former workshops of the Opera Garnier to the north of the site that are used to host various events. For the single industrial building—the recycling center—it was assumed to be fully maintained. Finally, the infrastructure was assumed to be maintained with the equivalent of 10 percent demolished during the transformation of the roads and parking lots next to the large Palais de Justice by the Piano Building Workshop. In this waste management phase, it was estimated that a little over 60 percent was recycled and most of the remainder resorted. Of the recycled materials 60 percent were assumed to be mined as mentioned earlier and among the sorted waste around 62 percent was landfilled, with only marginal amounts incinerated and composted.

Overall, 72 percent of the flows remained in building stock, with some 28 percent going to collected waste. This is a considerably larger apportionment to waste for the purposes of this illustration than in the more realistic contemporary material stock-flow model of the Clichy-Batignolles ZAC presents later on in the case study section, where 97 percent of the flow remains as it is today in the existing building stock and as one might expect for a relatively new redevelopment project.

G. Data Sources.

Data for the modeling exercises in the volume covered at least seven broad categories. They were spatial and population data, construction data, transportation data, information regarding specific stocks of material, and data about waste management and recycling, as well as about water and energy use. The types of sources involved were primarily governmental agencies, though also proprietary sources pertinent to particular kinds of materials, such as Boisie Cascade in the U.S. regarding timber products. Over time government sources appear to be improving, particularly under the auspices of anthropogenic concerns, like climate change and brownfield sites. Overall, there were three kinds of data, as depicted in Figure 5. The first used direct field observation and measurement. This usually concerned information about site uses, some population characteristics, and areas of coverage. Second, there was data specific to a particular locale and within the purview of that locale, such as a ZAC or special development zone in Paris or the Westwood Municipality in Boston. Standard governmental tract designations were also adhered to when possible such as census tracts and zip codes in the U.S. or similar in other national settings. Third, there was data relying on proxies drawn from wider surrounding areas such as the city-wide scale. Rarely, if ever, data sources went beyond these levels of spatial resolution. Rough proportions among each of these three categories were around 5 to 10 percent from direct observation, some 25 percent from local sources, and the remainder involving broader proxies. All data, however, was clearly classified as to source, documented, and replicable.

Data	Input	Existing condition with vehicles	Existing condition without vehicles
COMMERCIAL ACTIVITIES			
Commercial buildings area (m2)	72700	72700	72700
Number of hausmannian commercial buildings	62	62	62
Haussmannian commercial buildings area (m2)	70740	70740	70740
Number of post 1920 commercial buildings	3	3	3
Post 1920 commercial buildings area (m2)	1960	1960	1960
Retrofit rate	0.01	0.01	0.01
Restaurants and food shops/courts (m2)	13000	13000	13000
Retrofit rate	0.03	0	0
Bistro-type restaurants (m2)	5800	5800	5800
Sales food from plant (ton/m2/y)	1.25	1	1
Sales food from animal (ton/m2/y)	0.76	1	1
Sales beverage from water (ton/m2/y)	0.72	1	1
Sales packages from plastic (ton/m2/y)	0.1	0	0
Sales packages from paper (ton/m2/y)	0.04	0.04	0.04
Sales packages from glass (ton/m2/y)	0.09	0.09	0.09
Building furniture from plastic (ton/m2)	0.0202	0.0202	0.0202
Building furniture from wood (ton/m2)	0.0195	0	0
Building furniture from aluminum (ton/m2)	0.1484	0.1484	0.1484
Building furniture from iron (ton/m2)	0.0881	0.0881	0.0881
Building appliances from steel (ton/m2)	0.0879	0.0879	0.0879
Building appliances from plastic (ton/m2)	0.005	0.005	0.005
Building appliances from aluminum (ton/m2)	0.0223	0.0223	0.0223
Fast food restaurants (m2)	6000	6000	6000
Sales food from plant (ton/m2/y)	1.5	1.5	1.5
Sales food from animal (ton/m2/y)	0.64	0.64	0.64
Sales beverage from water (ton/m2/y)	0.61	0.61	0.61
Sales packages from plastic (ton/m2/y)	0.19	0.19	0.19
Sales packages from paper (ton/m2/y)	0.12	0.12	0.12
Sales packages from aluminum (ton/m2/y)	0.22	0.22	0.22
Building furniture from plastic (ton/m2)	0.0588	0.0588	0.0588
Building furniture from aluminum (ton/m2)	0.1113	0.1113	0.1113
Building appliances from steel (ton/m2)	0.2638	0.2638	0.2638
Building appliances from plastic (ton/m2)	0.0151	0.0151	0.0151
Building appliances from aluminum (ton/m2)	0.0668	0.0668	0.0668

Bakeries and assimilated (m2)	1200	1200	1200
Sales food from plant (ton/m2/y)	3.14	3.14	3.14
Sales food from animal (ton/m2/y)	0.33	0.33	0.33
Sales beverage from water (ton/m2/y)	0.62	0.62	0.62
Sales packages from plastic (ton/m2/y)	0.06	0.06	0.06
Sales packages from paper (ton/m2/y)	0.19	0.19	0.19
Sales packages from aluminum (ton/m2/y)	0.17	0.17	0.17
Building furniture from plastic (ton/m2/y)	0.0101	0.0101	0.0101
Building furniture from wood (ton/m2)	0.0097	0.0097	0.0097
Building furniture from aluminum (ton/m2)	0.0742	0.0742	0.0742
Building appliances from steel (ton/m2)	0.0879	0.0879	0.0879
Building appliances from plastic (ton/m2)	0.005	0.005	0.005
Building appliances from aluminum (ton/m2)	0.0223	0.0223	0.0223
Retail stores (m2)	34900	34900	34900
Retrofit rate	0.018	0.018	0.018
Clothes and accessories (m2)	18800	18800	18800
Sales clothes from plant (ton/m2/y)	0.5	0.5	0.5
Sales clothes from animal (ton/m2/y)	0.08	0.08	0.08
Sales clothes from plastic (ton/m2/y)	4.02	4.02	4.02
Sales clothes from other (ton/m2/y)	4.38	4.38	4.38
Sales packages from plastic (ton/m2/y)	0.03	0.03	0.03
Sales packages from paper (ton/m2/y)	0.03	0.03	0.03
Building furniture from plastic (ton/m2)	0.0168	0.0168	0.0168
Building furniture from wood (ton/m2)	0.0097	0.0097	0.0097
Building furniture from aluminum (ton/m2)	0.0445	0.0445	0.0445
Building appliances from steel (ton/m2)	0.0132	0.0132	0.0132
Building appliances from plastic (ton/m2)	0.0008	0.0008	0.0008
Building appliances from aluminum (ton/m2)	0.0033	0.0033	0.0033
Jewellery and luxury goods (m2)	3900	3900	3900
Sales accessories from animal (ton/m2/y)	0.63	0.63	0.63
Sales consumer products from glass (ton/m2/y)	0.52	0.52	0.52
Sales consumer products from other metals (ton/m2/y)	22.08	22.08	22.08

FIG.3: Three Kinds of Data Sources.

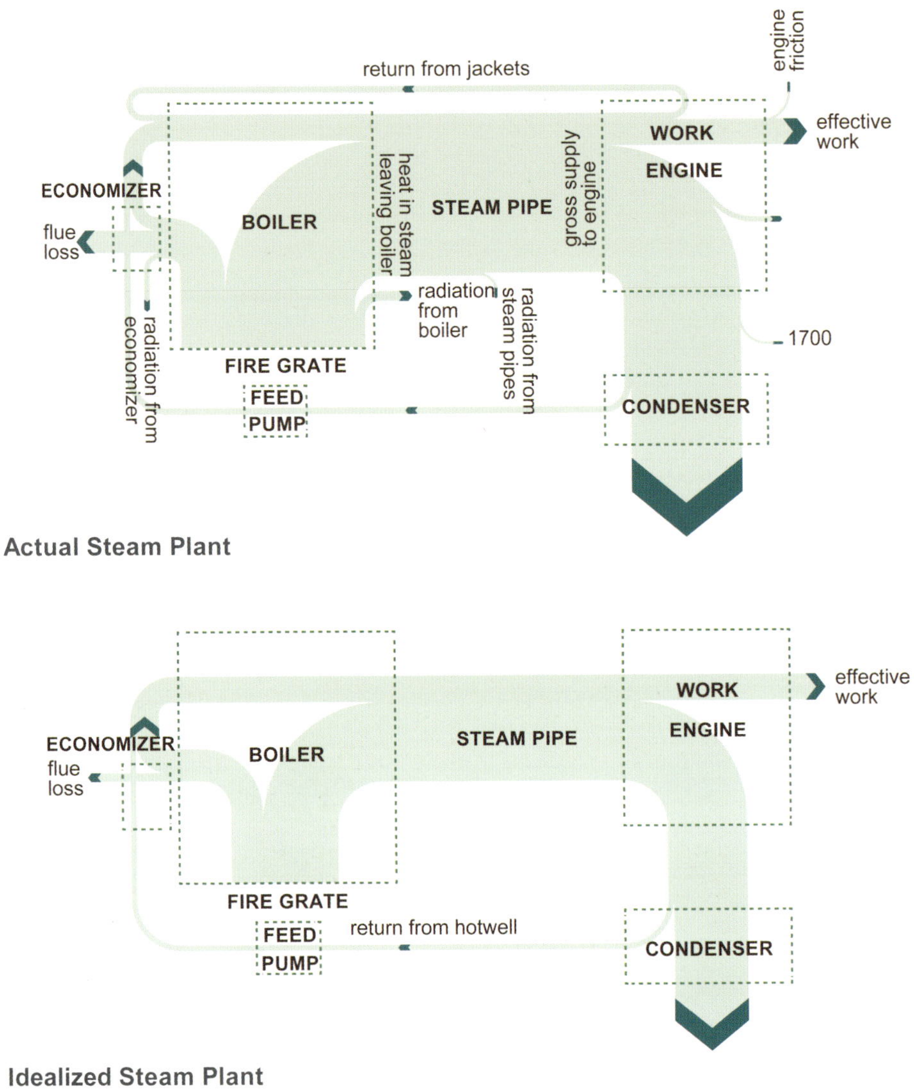

FIG.4: Sankey's Diagram of the Steam Engine.

H. Representation of Results.

Basic representation of data in the stock-flow modeling takes the form of Microsoft Excel spreadsheets, first introduced in 1987 and utilizing algebraic formulas for organizing and computing resulting values. Final results, however, are represented graphically with Sankey diagrams, named for the Irish Captain Sankey (1853–1926), a mechanical engineer and soldier, who used the diagram technique to show the efficiency of the steam engine in 1898 and as depicted in Figure 4. Flow depictions of data generally are grouped under alluvial diagrams named for their apparent graphic similarity to alluvial flows of soil deposition. The primary difference, though, is that Sankey diagrams show explicitly how quantities of specific elements flow from one state to another, as indicated in Figure 5. The proprietary software SankeyMATIC allows for relatively easy plotting of Sankey diagrams, though with the drawback of not representing circularity easily, which needed to be overcome through obsolete equations. In this volume, as noted earlier, the stage-processes of settlement development are the subjects of the diagrams, as depicted in Figure 2. Consequently, left-to-right positions in the Sankey diagrams correspond to stages of transition and transformation from 'cradle-to-grave' via end uses (Schmidt, 2008; and Peterson, 2020). Earlier versions of flow diagrams similar to Sankey diagrams were devised by Charles Joseph Minard (1781–1870), a French civil engineer, who used them in 1869 to describe, among other events, Napoleon's ruinous Russian campaign of 1812 (Rendgen, 2018). In summary, Sankey diagrams are a type of flow diagram in which the widths of the flow depictions are proportional to the flow rates of the depicted property in weight or volume over time. In the modeling exercises depicted here there are three basic types. They are water, energy, and material flows.

I. Model Interfaces.

Within each of the three types of stock-flow model depicted in this volume, special interfaces have been developed, particularly for facilitation of data input. In the water model and Figure 6, for instance, these include information about the site and occupying population; residential, commercial, and public; as well as

Estimated U.S. Energy Consumption in 2022 100.3 Quads

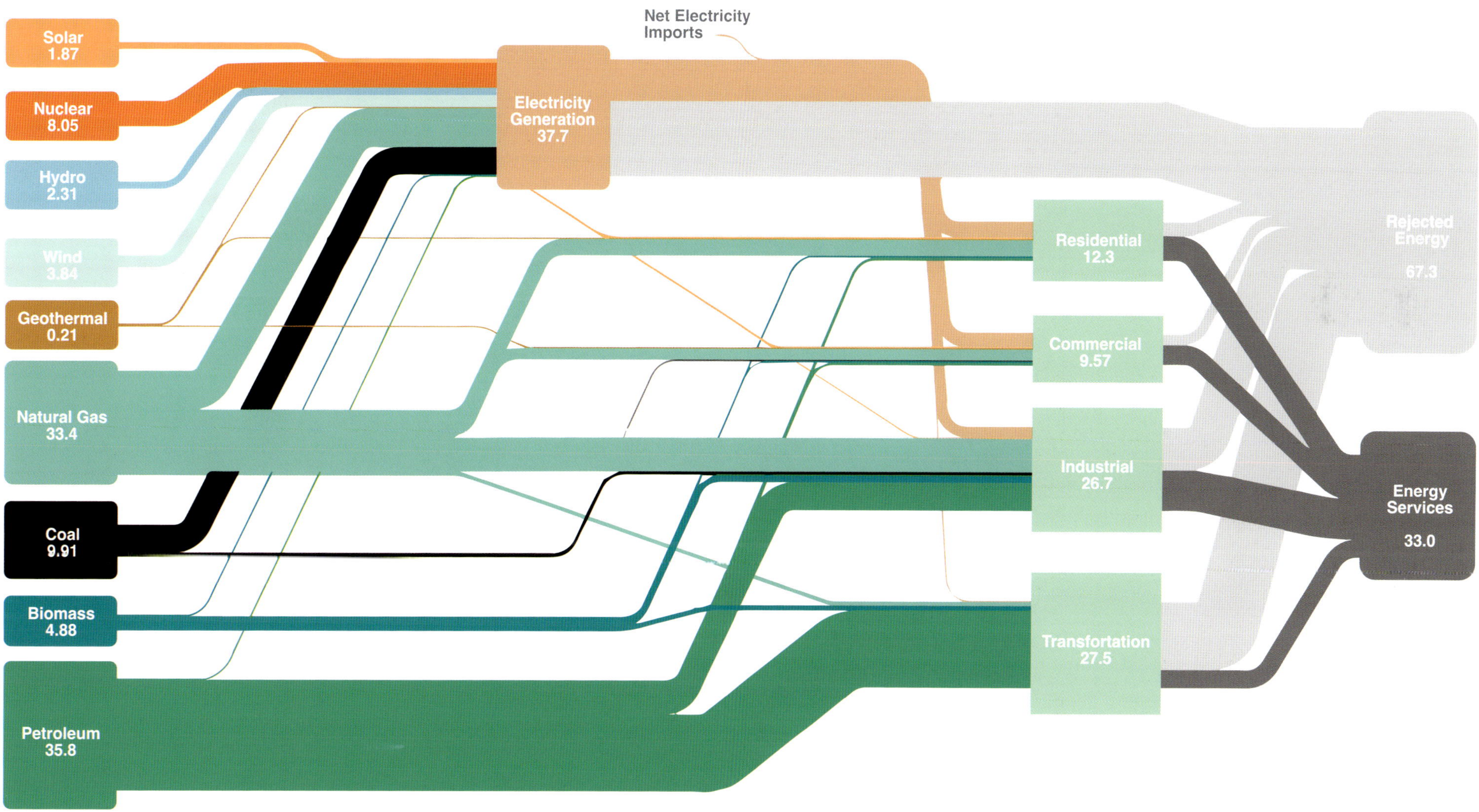

FIG. 5: A Sankey Diagram Depicting Estimated U.S. Energy Consumption.

industrial and agricultural activities, along with green spaces. The energy model interfaces apply again to site information and population characteristics, and commercial, industrial, agricultural, and transport characteristics, as shown in Figure 7. For the material models, categories include site and population; construction of buildings and infrastructure; green spaces; production from agriculture and industry; activities and goods consumption; transport characteristics like various fleets; and waste, including recycling, composting, mining, and waste disposal as depicted in Figure 8.

In summary, using the mass conservation principle, the metabolic model can be expressed as follows:

$$\Sigma p\ \dot{m}input = \Sigma q\ \dot{m}output + \dot{m}storage$$

Where, m = mass or other equivalent inputs for water and energy, and q = summation over inputs and outputs.
Mstorage = Stock + Collected Waste.
Collected Waste = Recycled + Un-reused Waste.

Throughout, essentially the models use an MFA (Mass Flow Analysis) scheme using flow data at a neighborhood and city level, as well as at higher levels of region or country as needed (Derrible, Cheah, Arora, and Yeow, 2021).

Data	Input
Total study area (m2)	
Rainfall (mm/year)	
% of water production from surface water	
% of water production from ground water	
% of water production from stormwater	
Residential	
Number of households	
Average water consumption of households (m3/y)	
% of toilet consumption (From drinking water)	
% of Laundry consumption (From drinking water)	
% of Dishwashing consumption (From drinking water)	
% of Shower/bath consumption (From drinking water)	
% of Faucet consumption (From drinking water)	
% of Outdoor consumption (From drinking water)	
% of Outdoor consumption (From storm water)	
% Leaks	
Commercial	
Commerce total area (m2)	
Average water consumption of commerces (m3/m2/y)	
% Leaks	
Industrial	
Industry total area (m2)	
Average water consumption of industries (m3/m2/y)	
% Leaks	
Agricultural	
Agriculture total area (m2)	
Average water consumption of agriculture (m3/m2/y)	
% Leaks	
Public Green Spaces	
All green permeable space subject to indirect watering (from stormwater) (m2)	
Green permeable space subject to direct watering (from drinking water) (m2)	
Total impermeable space (m2)	
Average watering consumption of greenspaces (from drinking water) (m3/m2/y)	
% watering consumption of greenspaces (from stormwater)	
% Leaks	

FIG.6: Water Model Input Interface.

Residential Activities	
Number of households	
Average energy consumption household (kwh/y)	
% Heating consumption	
% From electricity	
% From natural gas	
% From petroleum	
% From coal	
% From other	
% Losses heating	
% Cooling consumption	
% From electricity	
% From natural gas	
% From petroleum	
% From coal	
% From other	
% Losses cooling	
% Lighting and appliances consumption	
% From electricity	
% From Natural Gas	
% From Petroleum	
% From coal	
% From other	
% Losses lighting and appliances	
% Total losses	

Commercial Activities	
Floor area (m2)	
Average energy consumption (kwh/m2/y)	
% Heating consumption	
% From electricity	
% From natural gas	
% From petroleum	
% From coal	
% From other	
% Losses heating	
% Cooling consumption	
% From electricity	
% From natural gas	
% From petroleum	
% From coal	
% From other	
% Losses cooling	
% Lighting and appliances consumption	
% From electricity	
% From natural gas	
% From petroleum	
% From coal	
% From other	
% Losses lighting and appliances	
% Other	
% From electricity	
% From natural gas	
% From petroleum	
% From coal	
% From other	
% Losses other	
% Losses	

Data	Input
Electricity Production	
Renewable electricity production on site (kwh/y)	
% From renewables	
% From nuclear	
% From natural gas	
% From petroleum	
% From coal	
% Electricity production losses	
Industrial Activities	
Floor area (m2)	
Average energy consumption (kwh/m2/y)	
% From electricity	
% From natural gas	
% From petroleum	
% From coal	
% From other	
% Losses	
Agricultural Activities	
Floor area (m2)	
Average energy consumption (kwh/m2/y)	
% From electricity	
% From natural gas	
% From petroleum	
% From coal	
% From other	
% Losses	
Transport	
Number of km per household per year	
Number of kwh per km	
% From electricity	
% From petroleum	
% From natural gas	
% Losses	

FIG.7: Energy Model Input Interface.

Data	Input
Population	
Number of residents	
Number of households	
Construction	
Buildings	
Floor area collective housing haussmanian buildings and equivalent (m2)	
Building footprint housing haussmanian buildings and equivalent (m2)	
Number of collective housing haussmanian buildings and equivalent	
Floor area collective housing post-1920 buildings (m2)	
Floor area commercial and public haussmanian buildings and equivalent (m2)	
Building footprint commercial and public haussmanian buildings and equivalent (m2)	
Number of commercial and public haussmanian buildings and equivalent	
Floor area commercial and public post-1920 buildings (m2)	
Floor area industrial post-1920 building (m2)	
Floor area agriculture pre-1920 building (m2)	
Building footprint agriculture pre-1920 building (m2)	
Number of agriculture pre 1920 buildings	
Floor area agriculture post-1920 building (m2)	
Life cycle collective housing haussmanian buildings and equivalent (y)	
Life cycle collective housing post-1920 buildings (y)	
Life cycle commercial haussmanian buildings and equivalent (y)	
Life cycle commercial and public post-1920 buildings (y)	
Life cycle industrial post-1920 building (y)	
Life cycle agriculture pre-1920 building (y)	
Life cycle agriculture post-1920 building (y)	
Infrastructure	
Asphalt roads (m2)	
Parking lots (m2)	
Railways rails length (m)	
Life cycle asphalt roads (y)	
Life cycle parking lots (y)	
Life cycle railways rails (y)	
Green Spaces	
Leisure land area herbaceous strata (m2)	
Leisure land area shrub strata (m2)	
Leisure number of trees	
Life cycle trees (y)	
Life cycle shrub strata (y)	
Life cycle herbaceous strata (y)	
Production	
Agriculture	
Production land area herbaceous strata (m2)	
Production land area shrub strata (m2)	
Production number of trees	
Number of bovines	
Number of pigs, sheeps and equivalent	
Number of chicken and equivalent	
Organic waste (%)	
Industrial	
Produced plant (t/y)	
Produced animal (t/y)	
Produced aluminum (t/y)	
Produced iron (t/y)	
Produced steel (t/y)	
Produced wood (t/y)	
Produced paper (t/y)	
Produced plastic (t/y)	
Produced glass (t/y)	
Produced ruber (t/y)	
Activities and Goods	
Food and Drinks	
Average consumption of plant based food per capita (kg/y)	
Average consumption of animal based food per capita (kg/y)	
Average consumption of water per capita (kg/y)	
Organic food waste (%)	
Packages	
Average consumption of plastic in packages per capita (kg/y)	
Average consumption of paper in packages per capita (kg/y)	
Average consumption of glass in packages per capita (kg/y)	
Furniture and Furnishing	
Average weight of plastic in furniture per capita (kg)	
Average weight of wood in furniture per capita (kg)	
Life cycle furniture and furnishing (y)	
Consumer and Institutional Products	
Average weight of plastic in consumer and institutional products per capita (kg)	
Average weight of paper in consumer and institutional products per capita (kg)	
Average weight of glass in consumer and institutional products per capita (kg)	
Life cycle plastic consumer and institutional products (y)	
Life cycle paper consumer and institutional products (y)	
Life cycle glass consumer and institutional products (y)	
Clothing	
Average weight of plant products in clothing per capita (kg)	
Average weight of animal products in clothing per capita (kg)	
Average weight of plastic in clothing per capita (kg)	
Average weight of other products in clothing per capita (kg)	
Life cycle of clothing (y)	
Appliances	
Average consumption of steel in major appliances per household (kg)	
Average consumption of plastic in major appliances per household (kg)	
Average consumption of aluminum in major appliances per household (kg)	
Life cycle of appliances (y)	

FIG.8: Material Model Input Interface.

Transport

Private Cars Fleet

Number of cars	
Average weight of steel used per car (kg)	
Average weight of plastic used per car (kg)	
Average weight of aluminum used per car (kg)	
Average weight of rubber used per car (kg)	
Life cycle cars (y)	

Bus Fleet

Number of buses	
Average weight of steel used per bus (t)	
Average weight of plastic used per bus (t)	
Average weight of aluminum used per bus (t)	
Average weight of rubber used per bus (t)	
Life cycle buses (y)	

Metro Fleet

Number of metros	
Average weight of steel used per metro (t)	
Average weight of plastic used per metro (t)	
Average weight of aluminum used per metro (t)	
Average weight of rubber used per metro (t)	
Life cycle metros (y)	

Tram Fleet

Number of trams	
Average weight of steel used per tram (t)	
Average weight of plastic used per tram (t)	
Average weight of aluminum used per tram (t)	
Average weight of rubber used per tram (t)	
Life cycle trams (y)	

Transition

% of maintenance residential	
% of repurposed residential	
% of demolished residential	
% of maintenance commercial and public	
% of repurposed commercial and public	
% of demolished commercial and public	
% of maintenance industrial	
% of repurposed industrial	
% of demolished industrial	
% of maintenance agricultural	
% of repurposed agricultural	
% of demolished agricultural	
% of maintenance infrastructure	
% of demolished infrastructure	
% of waste produced in repurposed compared to stock	
Maintenance condition residential (H=1 ; M=0,9 ; L=0,8)	
Maintenance condition commercial and public (H=1 ; M=0,9 ; L=0,8)	
Maintenance condition industrial (H=1 ; M=0,9 ; L=0,8)	
Maintenance condition agriculture (H=1 ; M=0,9 ; L=0,8)	
Maintenance condition infrastructure (H=1 ; M=0,9 ; L=0,8)	

Waste

Recycling rates (for mined and not mined)

% paper	
% plastic	
% aluminum	
% iron	
% steel	
% limestone	
% concrete	
% asphalt	
% wood	
% glass	
% clothing	
% rubber	

Industrial collected waste rates

% industrial waste from organic material (plant + animal)	
% industrial waste from aluminum	
% industrial waste from iron	
% industrial waste from steel	
% industrial waste from wood	
% industrial waste from paper	
% industrial waste from plastic	
% industrial waste from glass	
% industrial waste from ruber	

Composting rates

% organic waste food	
% organic waste greenspaces	
% organic waste agriculture	

Incinerated rates after sorting

% paper	
% plastic	
% aluminum	
% wood	
% clothing	
% rubber	

Landfill rates after sorting

% paper	
% plastic	
% aluminum	
% steel	
% iron	
% concrete	
% limestone	
% asphalt	
% wood	
% clothing	
% rubber	

Export rates after sorting

% steel	
% iron	
% concrete	
% limetone	
% asphalt	

04

Case Studies

The following chapter is comprised of representative case studies from the four domains of settlement coexistence, as well as several specific urban projects. The aim is to present results with regard to water flows, energy flows, and material flows, as well as to contextualize each case study area more fully with regard to spatial and population characteristics, locations within cities, and the physical character of settlement, including building and land-use types, infrastructure, and open space.

A. Specific Domains of Settlement Co-existence.

i. Compact City.

The representative site for the 'compact city' category is located in the Batignolles neighborhood, within the 17th Arrondissement of Paris in the north-west on the edge of the Boulevard Périphérique that circumscribes the city. Batignolles has about 100 or so Haussmann perimeter block buildings, around six stories in height and containing a mixture of uses, though mainly residential apartments on the order of 70 square meters in size with commercial and retail functions located on lower floors. The overall area is also generally well served by the Paris metro system and is relatively high in residential density and commercial intensity as depicted in Figure 1, more or less centered on the Boulevard Périphérique. The actual study area in this volume is the Clichy-Batignolles ZAC or '*Zone d'Aménagement Concerté*,' or Special Development Zone, that was set aside around 2002 as a site for innovation in environmental, affordable, and mixed-use development (Smurak, 2018). It is comprised of around 54

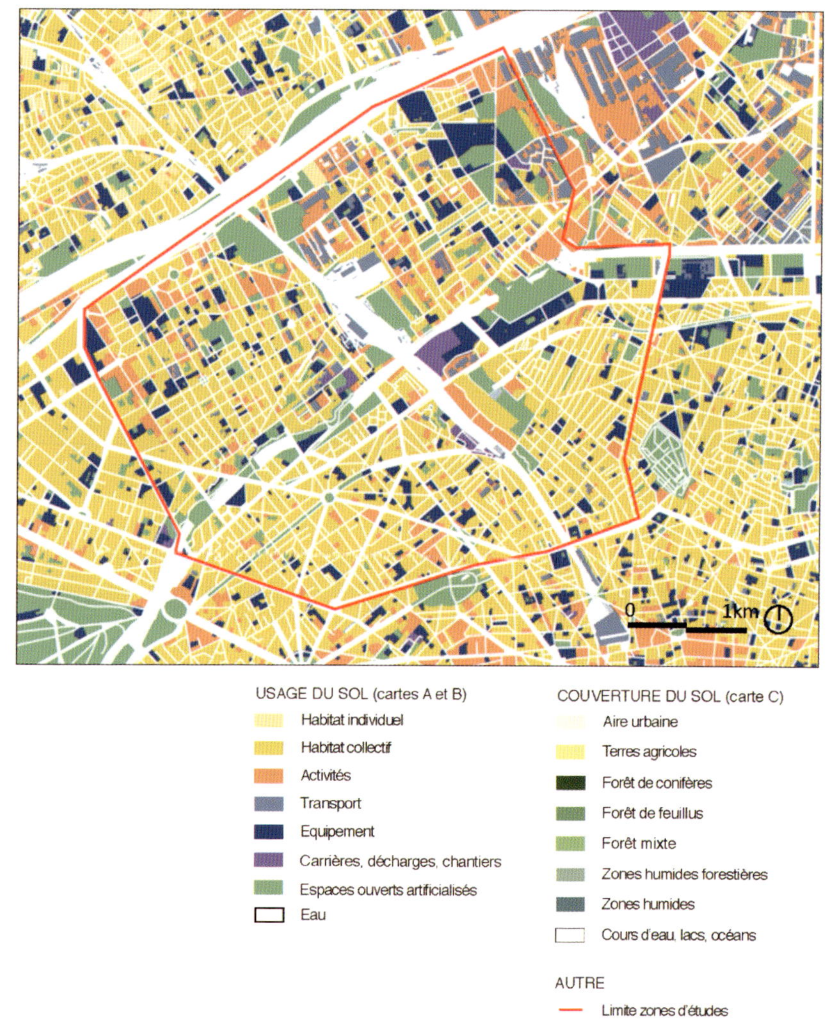

FIG. 1: The ZAC and the General Area of Clichy-Batignolles, Paris, France.

hectares of what was once an SNCF railroad yard as shown in Figure 3, on the border between the older Haussmann Paris of the 19th century and the more contemporary peripheral banlieue. Figure 2 also depicts the area of the ZAC within the more general area of Batignolles, along with the pre-ZAC land-use condition in Figure 3. The project was built to give a glimpse of a low-carbon future that so many climate plans envision, including those in France. As shown in Figure 4, in a plan coordinated by Francois Grether, it was comprised of the 10-hectare Martin Luther King Park designed by Jacqueline Osty, which also marks the gate to the commune of Clichy and, of course, commemorates the American Civil Rights Movement.

FIG.2: The General Area of Clichy-Batignolles, Paris, France on the Avenue Peripherique.

FIG. 3: Before and After Views of the ZAC at Clichy-Batignolles, Paris, France.
FIG. 4: The Plan of the Clichy-Batignolles ZAC by Francois Grether of 2002.

The overall plan accommodates around 7,500 inhabitants in 2,175 residences at around 3.4 inhabitants per unit, images of which are shown in Figure 5. In addition to residences there is also some 329,000 square meters of commercial space and around 9,800 square meters of industrial use. In addition, the new Ministry of Justice headquarters building complex by the Renzo Piano Building Workshop is located adjacent to the park and to the north of the Boulevard Périphérique, as shown in Figures 3 and 4. The Clichy-Batignolles ZAC deployed several technological strategies to achieve its environmental objectives as illustrated in Figure 6. Most of the heating was provided through a project-wide geothermal system. Some 35,000 square meters of solar photovoltaic panels affixed to buildings also offset off-site energy use substantially. In addition, the literal greening of buildings acted as an insulation, as well as hosting higher levels of biodiversity. Vacuum tubes conveyed waste from buildings to a site within the project where it was compacted to reduce the volume of truck transport needed to convey it away. Martin Luther King Park was replete with stormwater harvesting and management devices. Biodiversity was also enhanced through a rich array of planting in the park, as well as within numerous intermediate scales of public open space within the community. The mixing of uses was also provided for, along with provision of housing expressly for students and lower-income residents.

a. **Water Model.**

The defining characteristics of the water stock-flow model for the 'compact city' at Clichy Batignolles, Paris, and in particular the ZAC, and in cubic meters per year are as follows. As shown in figure Figure 7 the percentage of sources are surface water at around 23 percent, ground water at 23 percent, and stormwater at 54 percent from the sophisticated stormwater harvesting system on site mentioned earlier. The percentage distribution by use is residential at about 39 percent, community-public at 36 percent, industrial at around 1 percent, and greenspace at 24 percent. The percentage distribution of output by type is 59 percent to wastewater treatment, 10 percent in leakage, 18 percent ton evaporation, and 13 percent in infiltration. The total volume for the ZAC is 509,727 cubic meters per year.

FIG.5: New Buildings Within the Clichy-Batignolles ZAC in Paris, France.

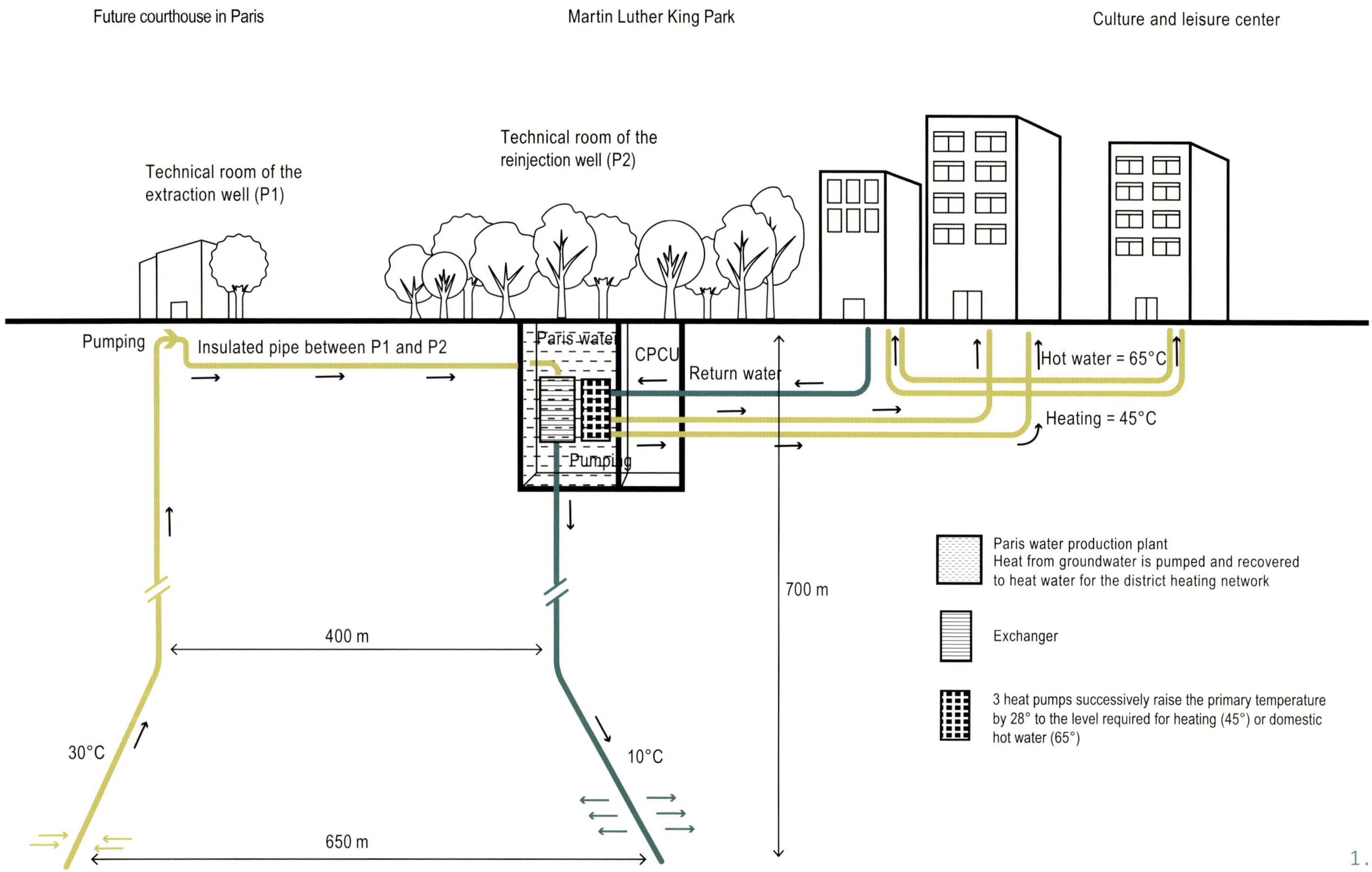

FIG.6: Some of the Ecological Management Components of the Clichy-Batignolles ZAC.

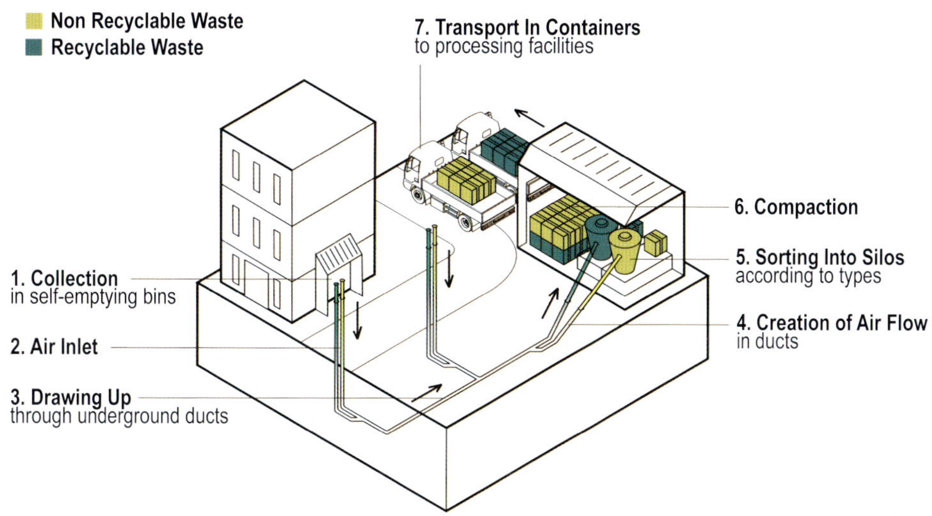
Non Recyclable Waste
Recyclable Waste
7. Transport In Containers
to processing facilities
6. Compaction
5. Sorting Into Silos
according to types
4. Creation of Air Flow
in ducts
1. Collection
in self-emptying bins
2. Air Inlet
3. Drawing Up
through underground ducts

2.

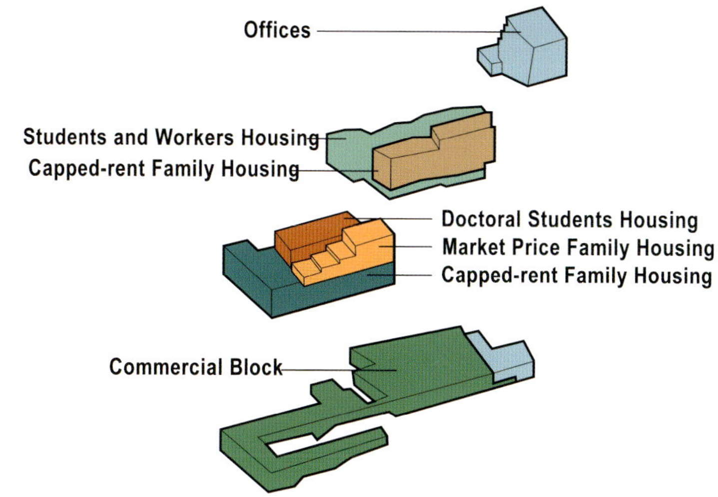
Offices
Students and Workers Housing
Capped-rent Family Housing
Doctoral Students Housing
Market Price Family Housing
Capped-rent Family Housing
Commercial Block

3.

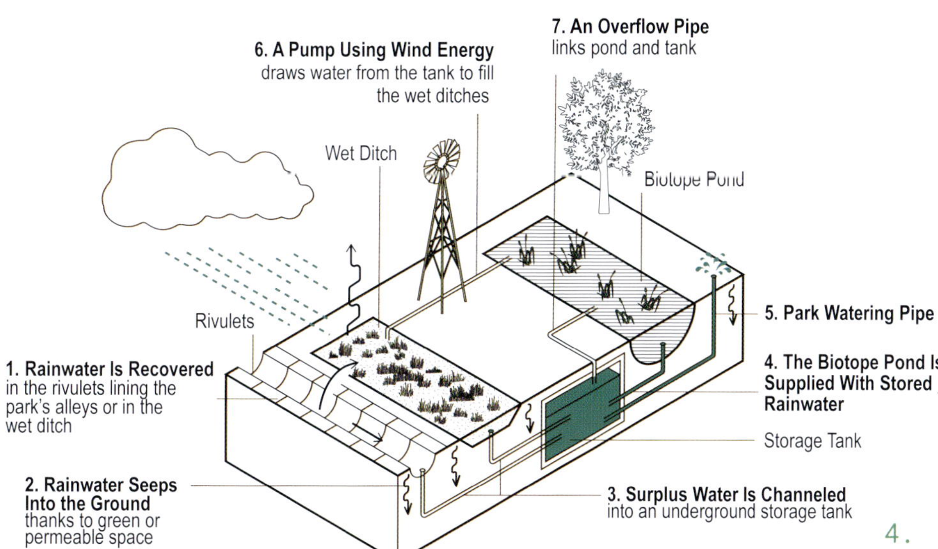
6. A Pump Using Wind Energy
draws water from the tank to fill
the wet ditches
7. An Overflow Pipe
links pond and tank
Wet Ditch
Rivulets
5. Park Watering Pipe
1. Rainwater Is Recovered
in the rivulets lining the
park's alleys or in the
wet ditch
4. The Biotope Pond Is
Supplied With Stored
Rainwater
Storage Tank
2. Rainwater Seeps
Into the Ground
thanks to green or
permeable space
3. Surplus Water Is Channeled
into an underground storage tank

4.

5.

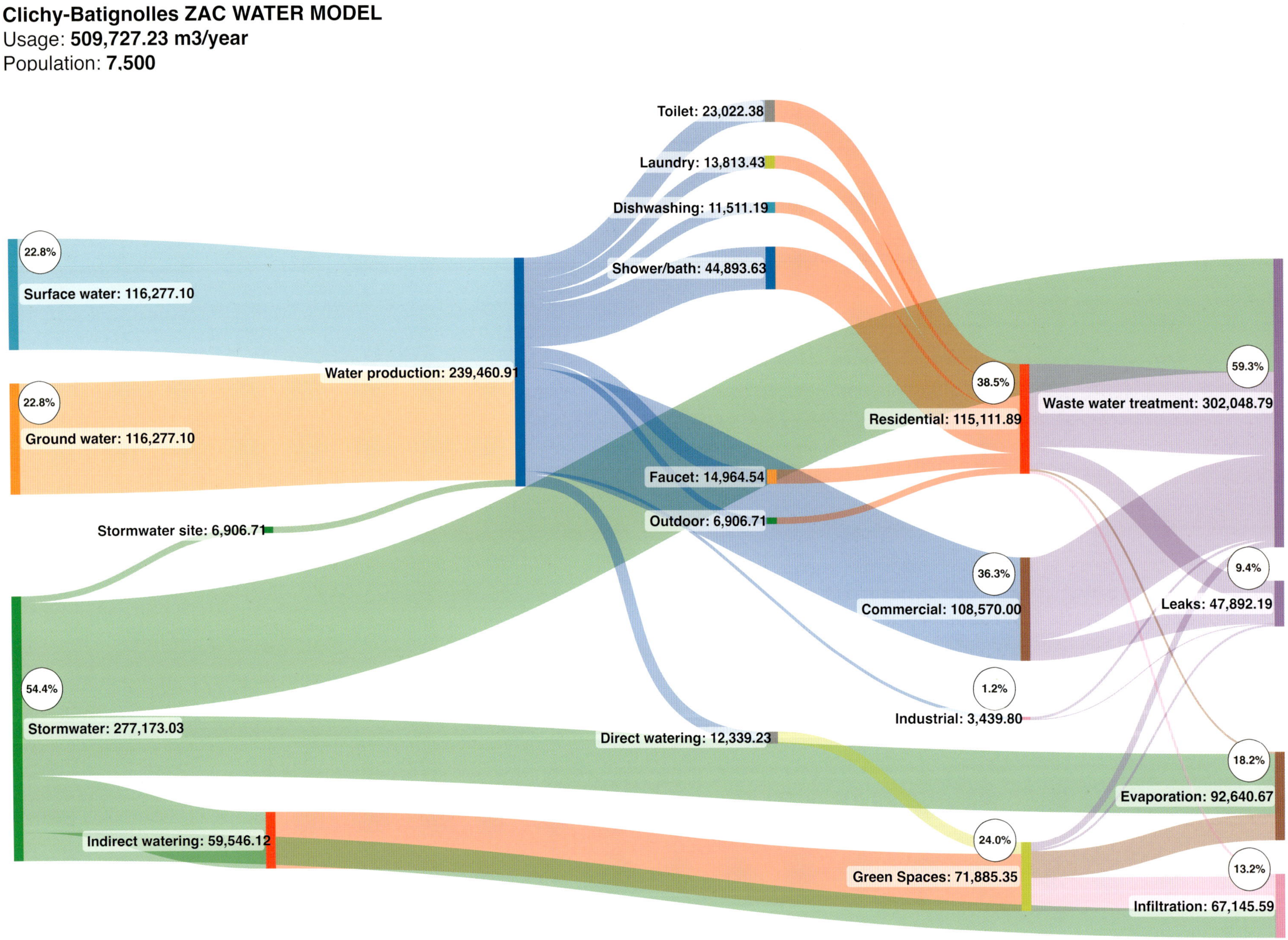

FIG.7: Paris ZAC Water Sankey Diagram.

b. **Energy Model.**

The defining characteristics of energy stock-flow model for the 'compact city' at Clichy Batignolles, Paris, and the ZAC there in particular in KWh/year are as follows. As shown in Figure 8 the percentage distribution by source is around 16 percent renewables, 59 percent nuclear, 25 percent fossil fuels including natural gas and petroleum, and 1 percent other. The percentage distribution by land use is 25 percent residential, 4 percent industrial, 33 percent commercial-public, and 38 percent transportation. The percentage distribution by output is about 62 percent lost and 38 percent consumed.

c. **Material Model**

The defining characteristics of the material stock-flow model in tons per year for the 'compact city' of the Clichy Batignolles ZAC in particular are as follows. As shown in Figure 9, 73 percent is extracted from the geosphere, 11 percent from the biosphere, and 16 percent from the hydrosphere. With regard to land uses, about 50 percent is residential, 44 percent commercial-public, 1 percent industrial, 2 percent infrastructural, 1 percent greenspace, and 2 percent transportation. With regard to waste management, recycling, and so on, 97 percent remains in existing stock, 2 percent on collected waste, and 1 percent is consumed. Then of the collected waste about 55 percent is recycled, 44 percent is sorted, and 1 percent is composted. The total material weight is 876,049 tons per year.

d. **Embodied Energy and Other Characteristics.**

The attached figure details the embodied energy in MJ/year and embodied carbon in kg CO_2/year for the ZAC at Clichy Batignolles. The average is 3.2 MJ/year of embodied energy and 0.28 kg CO_2/year of embodied carbon.

FIG.8: Paris ZAC Energy Sankey Diagram.

Clichy-Batignolles ZAC ENERGY MODEL
Usage: **159,577,908.41 kWh/year**
Population: **7,500**

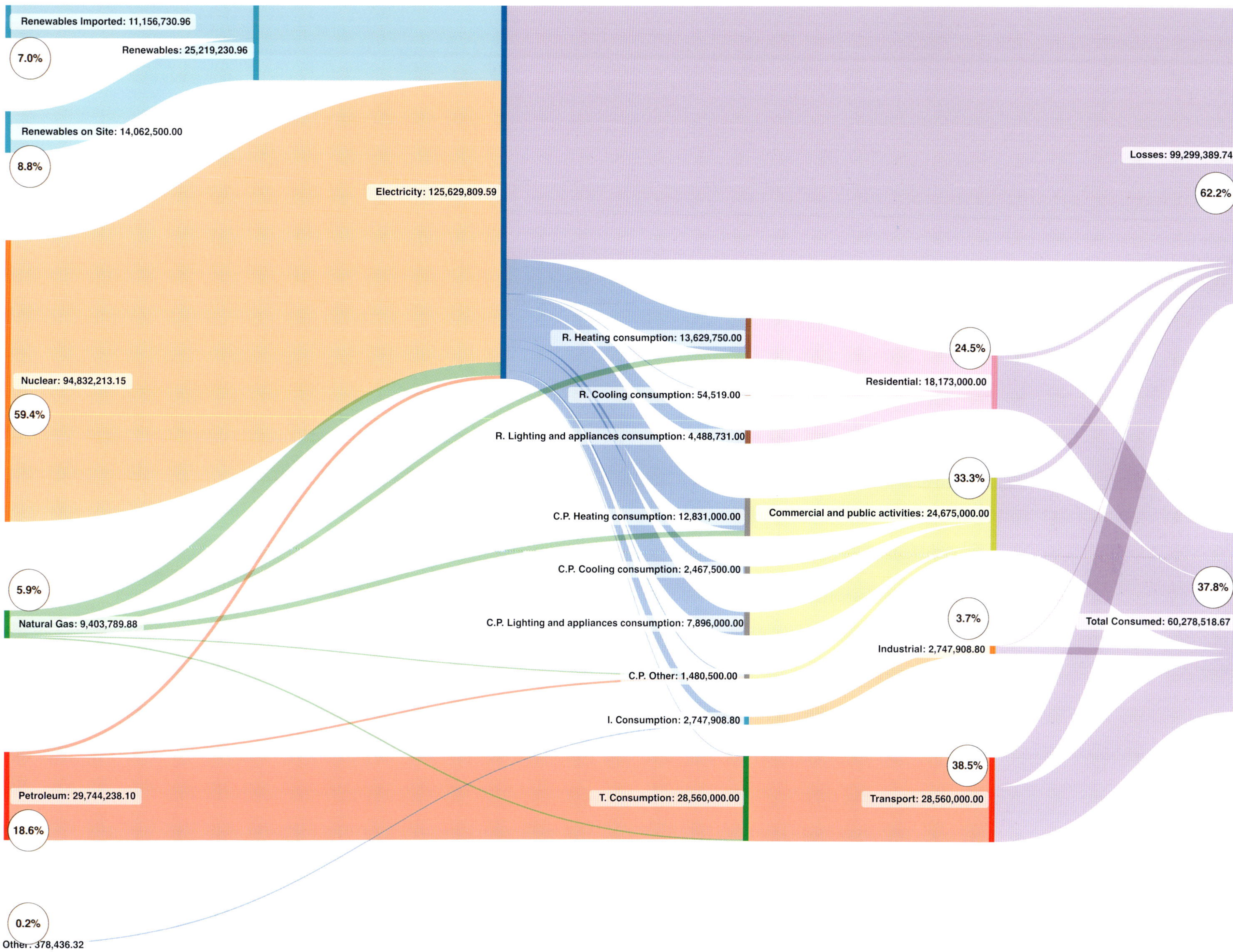

Clichy-Batignolles ZAC MATERIAL MODEL

Usage: **876,049.14 ton/year**
876049140 kg/year

Population: **7,500**

Lifecycle of buildings: 91 years
Lifecycle of roads: 15 years
Lifecycle of cars: 9.1 years
Lifecycle of metro: 15 years

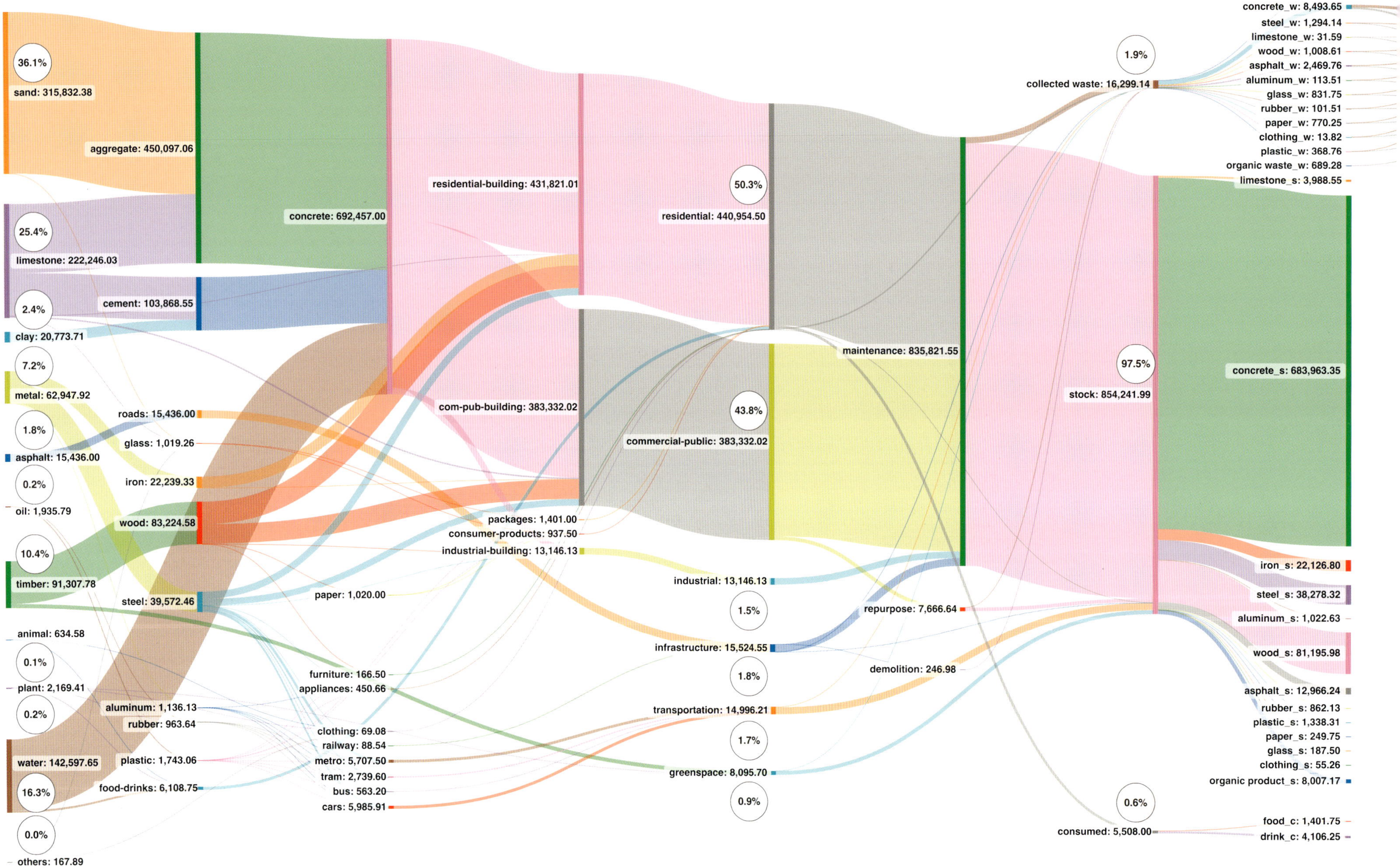

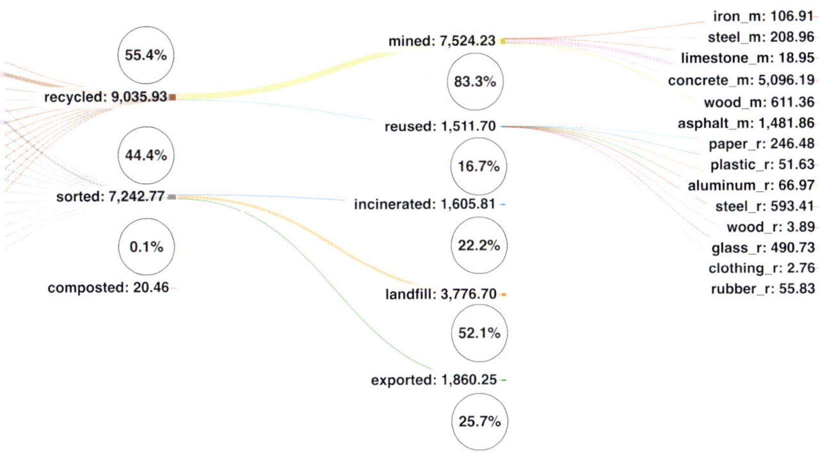

ZAC CLICHY-BATIGNOLLES

	Material	Ton/year (t/year)	Embodied energy (MJ/kg)	Embodied carbon (kgCO2/kg)	Total embodied energy (MJ/year)	Total embodied carbon (kgCO2/year)
1	Aggregate	450097	0.08	0.005	36007.76	2250.49
2	Limestone	222246	1	0.1	222246	22224.6
3	Clay	20774	3	0.23	62322	4778.02
4	Glass	1019	15.9	0.9 1	6202.1	917.1
5	Iron	22239	30	2.5	667170	55597.5
6	Timber	91308	0.3	0.3	27392.4	27392.4
7	Steel	39572	32	2.7	1266304	106844.4
8	Asphalt	15436	2.9	0.06	44764.4	926.16
9	Aluminum	1136	154	8.2	174944	9315.2
10	Rubber	964	110	6	106040	5784
11	Plastic	1743	95.4	4.98	166282.2	8680.14
	Grand Total	**866534**			**2789675**	**244710**
					MJ/kg/year Average 3.22	**kgCO2/kg/year Average 0.28**

FIG.9: Paris ZAC Material Sankey Diagram.
FIG.10: Matrix of Materials by Embodied Energy, etc for Paris ZAC.

ii. Dispersed City.

The representative site of the 'dispersed city' or peripheral development was in Norfolk County within the Boston Metropolitan Area of the U.S. As depicted in Figure 11 it is centered on Route 128, a circumferential state highway around Boston that was originally designated in 1927, some 92 kilometers in length and one of two beltways, the other being Interstate Highway 495, built as a part of the Federal Interstate Highway System, dating from World War II (Rowe, 1991). Locally it is generally recognized as the demarcation between the more urban inner suburbs and the lesser densely developed suburbs surrounding Boston. During the 1960s and 1980s it was the site of the location of high-technology firms in search of employees, particularly at the highly accessible junctions with major roads radiating out from the center of Boston. Indeed, the study site is centered on such an intersection of infrastructural elements, including the interstate AMTRAK rail line from Boston to New York City and beyond (Earls, 2002). Figure 11 also shows the relative drive times that gave rise to the patterns of accessibility that took the high-tech firms to these peripheral locations. In these regards the site is typical of American peripheral forms of urban development in metropolitan areas.

As shown on Figures 12, 13, and 14, this form of development is usually comprised of commercial and other non-residential development close to the freeway system and surrounded by or adjacent to low-density residential housing estates, many of which are subdivided into single-family homes. The actual site context is depicted in Figure 14 and is made up of zip codes in the townships of Westwood, Dedham, and Norwood, simply referred to here as Westwood. The combined population is 7,448 inhabitants, accommodated in 2,773 housing units, primarily single-family houses, spread across some 9.6 square kilometers or 962.5 hectares and at around 273 square meters on average per dwelling unit. The overall living density is quite low at around only 7.7 persons per gross hectare. Primarily in single-family homes, as shown in Figure 15, are some 380,000 square meters total. The amount of non-residential built space is around 540,000 square meters largely in commercial and public uses. Again, as depicted in Figure 13 and 14, it is located in clumps closely associated with

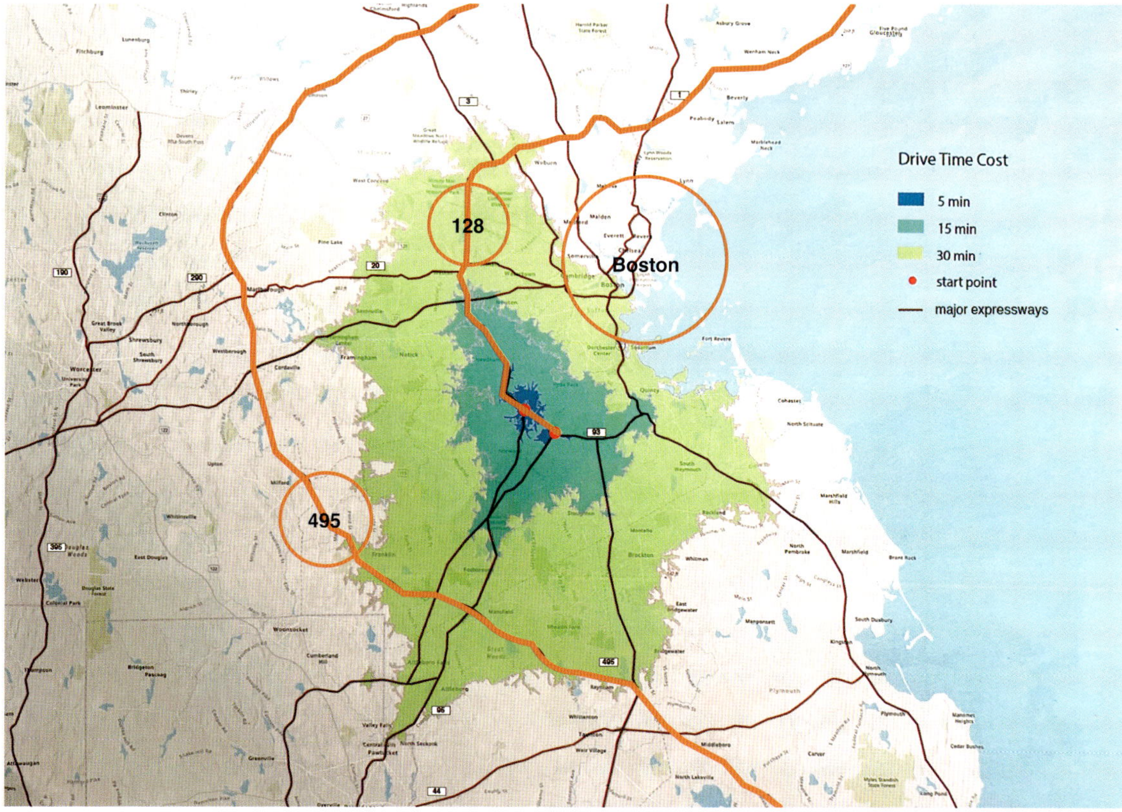

FIG.11: The Westwood Site in the Context of the Boston Metropolitan Area.
FIG.12: The Westwood Site in the Context of the Neponset Watershed.

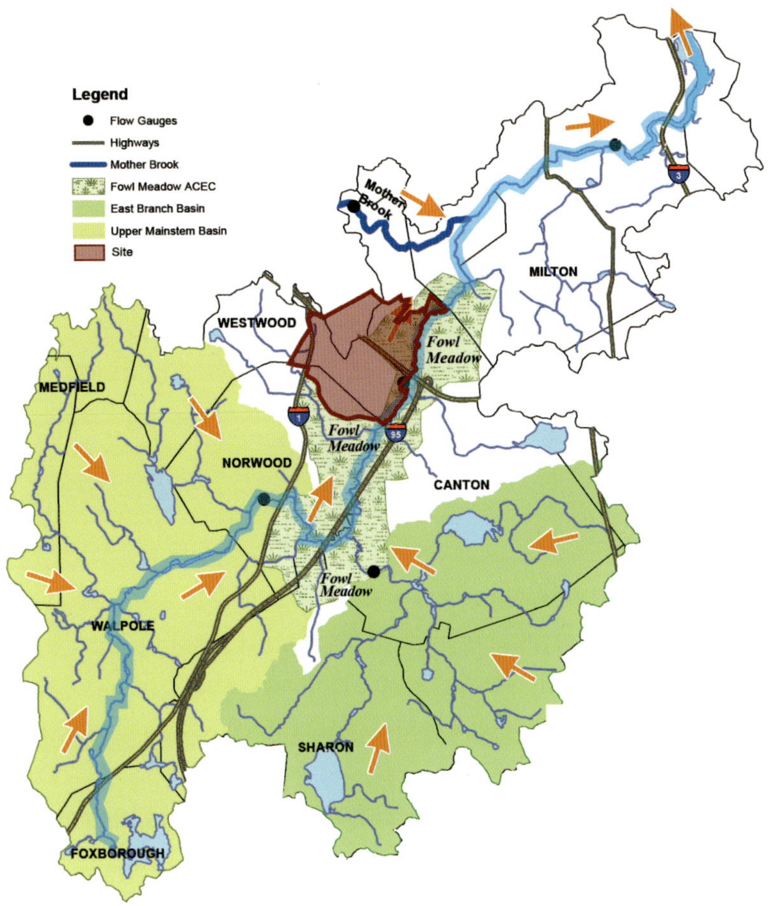

FIG. 13: The Westwood Site Within the Ecological Context of the Neponset Watershed.

major roads. Environmentally the study site is located mainly in the Neponset River watershed, with the river eventually spilling into the sea in central Boston at Dorchester Bay. Particularly as shown in Figure 13, the developed area abuts significant ecological preserves, such as the Fowl Meadow wetland system, as well as regulated natural zones largely by the Neponset River Watershed Association. In fact, Fowl Meadow together with nearby Blue Hills provide the most extensive area of wildlife habitat in the Greater Boston Area. In short, the watershed is a striking array of landscapes, habitats, and species refuges along the 48-or-so kilometer length of the river and within the study area amounting to some 68.7 percent of the land footprint area (Neponset River Watershed Association, 2017).

a. **Water Model**

The defining characteristics of the water stock-flow model for the 'dispersed city' at Westwood, Boston, in cubic meters per year and as shown in Figure 16, are as follows. The percentage distribution of sources is 10 percent surface and 90 percent groundwater. The percentage distribution by use is residential at around 36 percent, commercial-public at 45 percent, industrial at 1 percent, and maintained greenspace including private gardens at 18 percent. The percentage distribution in output by type is 64 percent wastewater treatment, 5 percent in leakage, 14 percent from evaporation, and 17 percent from infiltration. The improved or more efficient water model depicted in Figure 17 uses around 23 percent less water with some 57 percent of gray water from water treatment recycled into consumptive use. There is no stormwater harvesting employed at Westwood.

b. **Energy Model**

The defining characteristics of the energy stock-flow model for the 'dispersed city' at Westwood in KWh/year are as follows. As shown in Figure 18 the percentage distribution by source is about 21 percent in renewables, 77 percent in fossil fuels, and 2 percent other. The percentage distribution by land use is 23 percent residential, 39 percent commercial-public, 36 percent in transportation, and 2 percent in industrial. The percentage

Zipcode:		Population:		Housing Units:		Area:	
Westwood -	02090	Westwood -	3766	Westwood -	1429	Westwood -	4360803.96 sq m
Dedham -	02026	Dedham -	3667	Dedham -	1343	Dedham -	3992040.78 sq m
Norwood -	02062	Norwood -	15	Norwood -	1	Norwood -	1272301.51 sq m
		Total	**7448**	**Total**	**2773**	**Total**	**9625146.25 sq m**

CONTEXT MAP - WESTWOOD, DEDHAM AND NORWOOD

0 0.25 0.5 1 Miles N

FIG.14: The Context Map of Westwood, Dedham and Norwood.

FIG.15: Images of Residential Buildings in Westwood Site.

distribution by output is about 62 percent lost and 38 percent consumed. For the improved or more efficient energy model, as shown in Figure 19, use is made of electric vehicles, on-site solar arrays and heat pumps, as well as extra units sharing existing lots, fossil fuels are reduced by 76 percent and the ratio of energy lost to consumed improves by 79 percent.

c. **Materials Model**

The defining characteristics of the material stock-flow model in metric tons per year for the 'dispersed city' at Westwood, Boston, are as follows and are shown in Figure 20. A full 75 percent is drawn from the geosphere, 22 percent from the biosphere, and 3 percent from the hydrosphere. With regard to land uses, 46 percent is in paved infrastructural areas, 22 percent is commercial-public, 20 percent is residential, 1 percent is industrial, 8 percent is public greenspace, and 2 percent is on transportation. With regard to waste management, 96 percent remains in stock, almost 4 percent is collected waste, and the remainder is consumed. Of the collected waste, 25 percent is recycled, 69 percent is sorted, and 6 percent is composted. The total material weight is on the order of 1,838,209 tons per year, including non-building aspects of development.

d. **Embodied Energy and Other Characteristics.**

The attached figure details the embodied energy in MJ/year and the embodied carbon in kg CO2/year for Westwood. The average is 5.8 MJ/year of embodied energy and 0.32 kg CO2/year of embodied carbon.

WESTWOOD WATER MODEL
Usage: **2,955,954 m3/year**
Population: **7448**

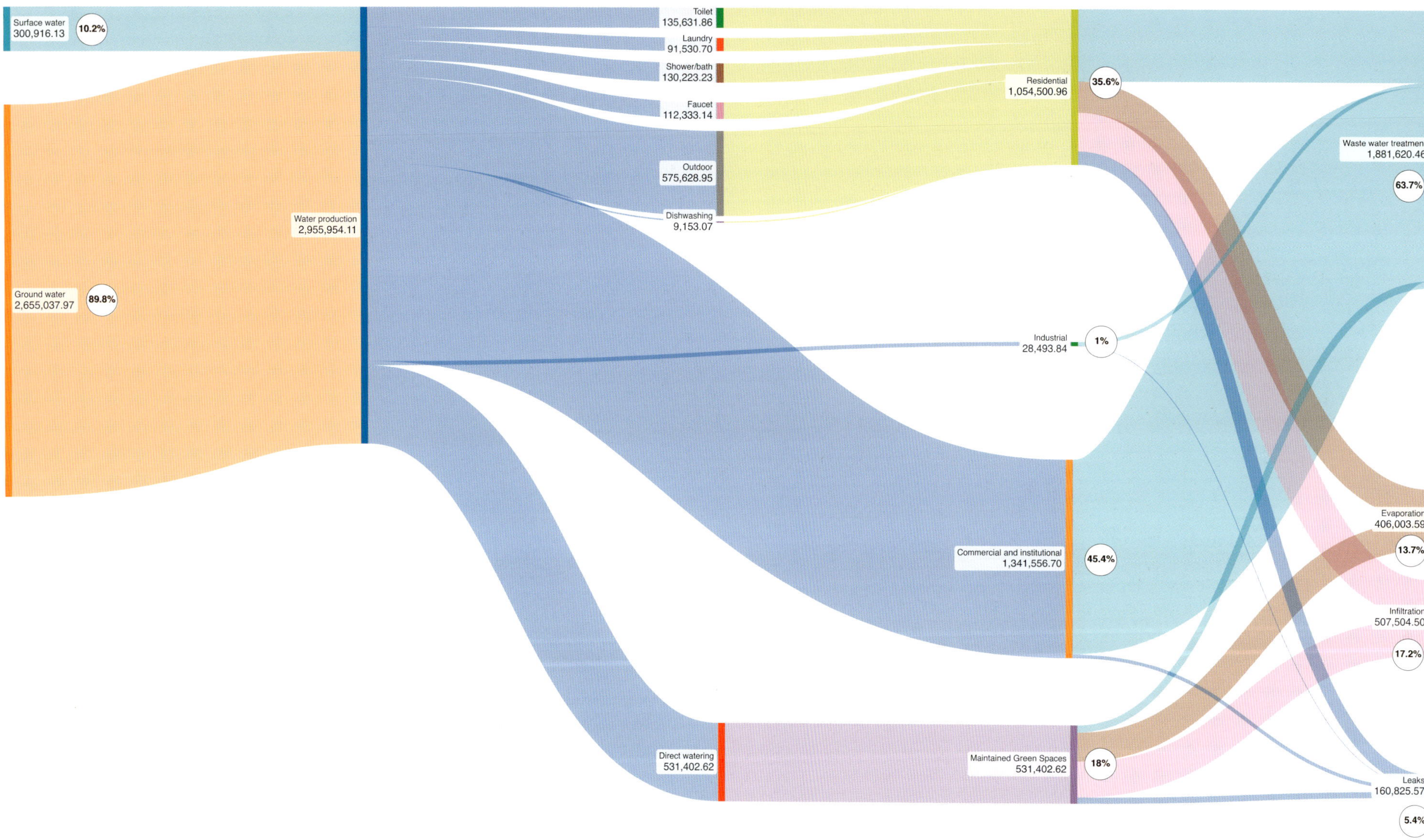

FIG.16: Westwood Water Sankey Diagram.

WESTWOOD EFFICIENT WATER MODEL

Usage: **2,262,565 m3/year**

Population: **7448**

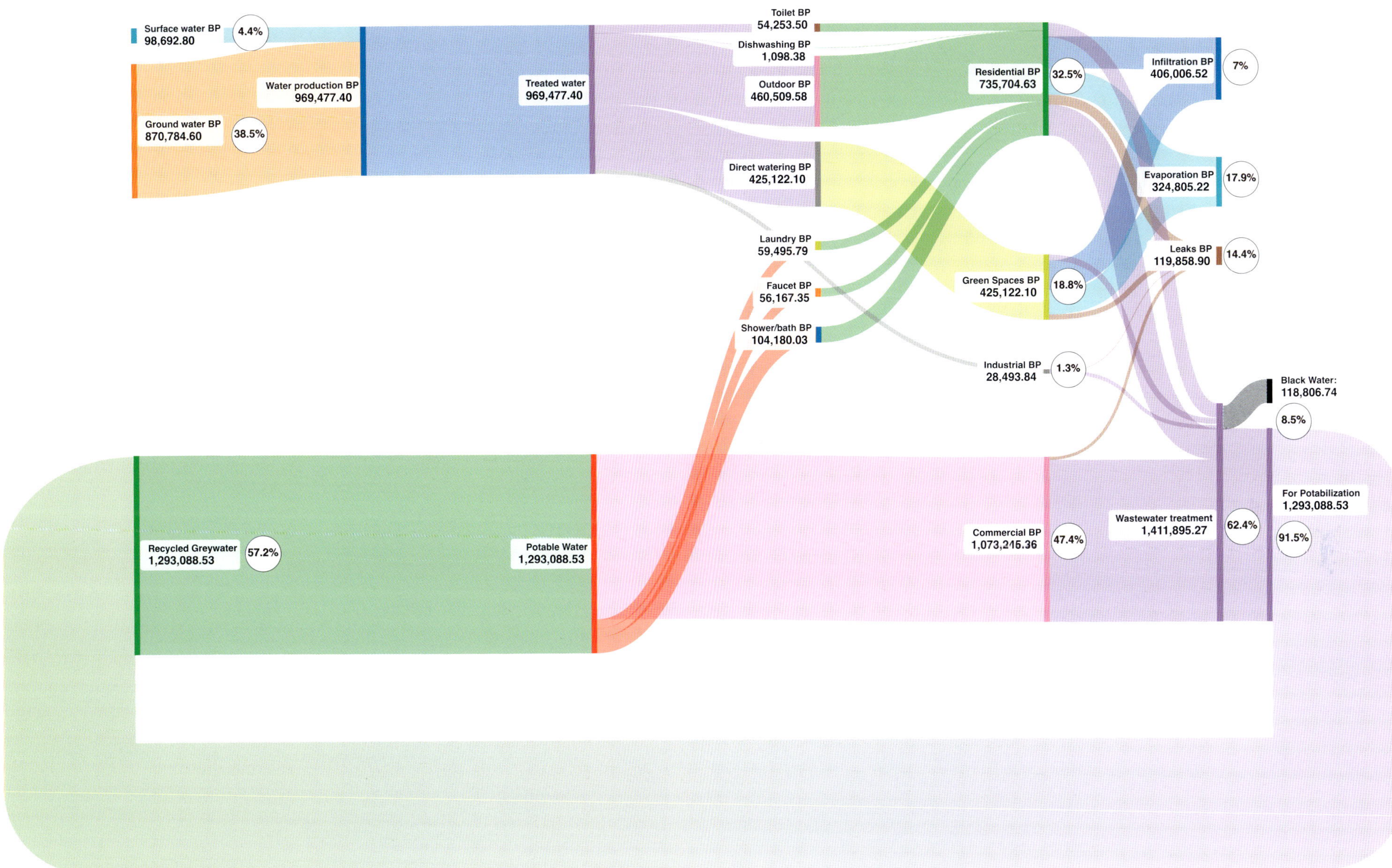

FIG.17: Westwood Water Efficiency Sankey Diagram.

WESTWOOD ENERGY MODEL

Usage: **480,085,877 kWh/year**
Population: **7448**

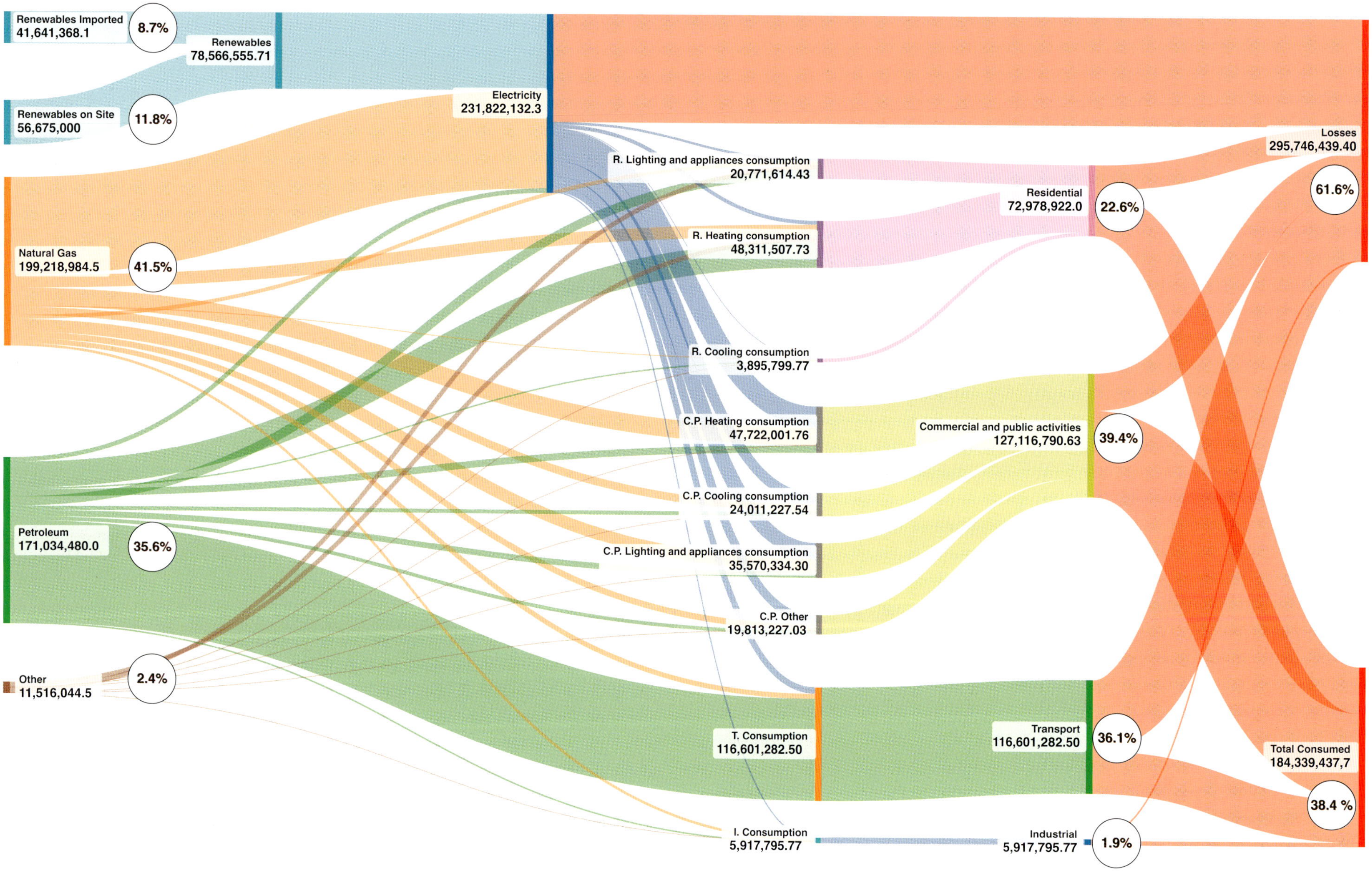

FIG.18: Westwood Energy Sankey Diagram.

WESTWOOD EFFICIENT ENERGY MODEL
Usage: **418,678,423 kWh/year**
Population: **7448**

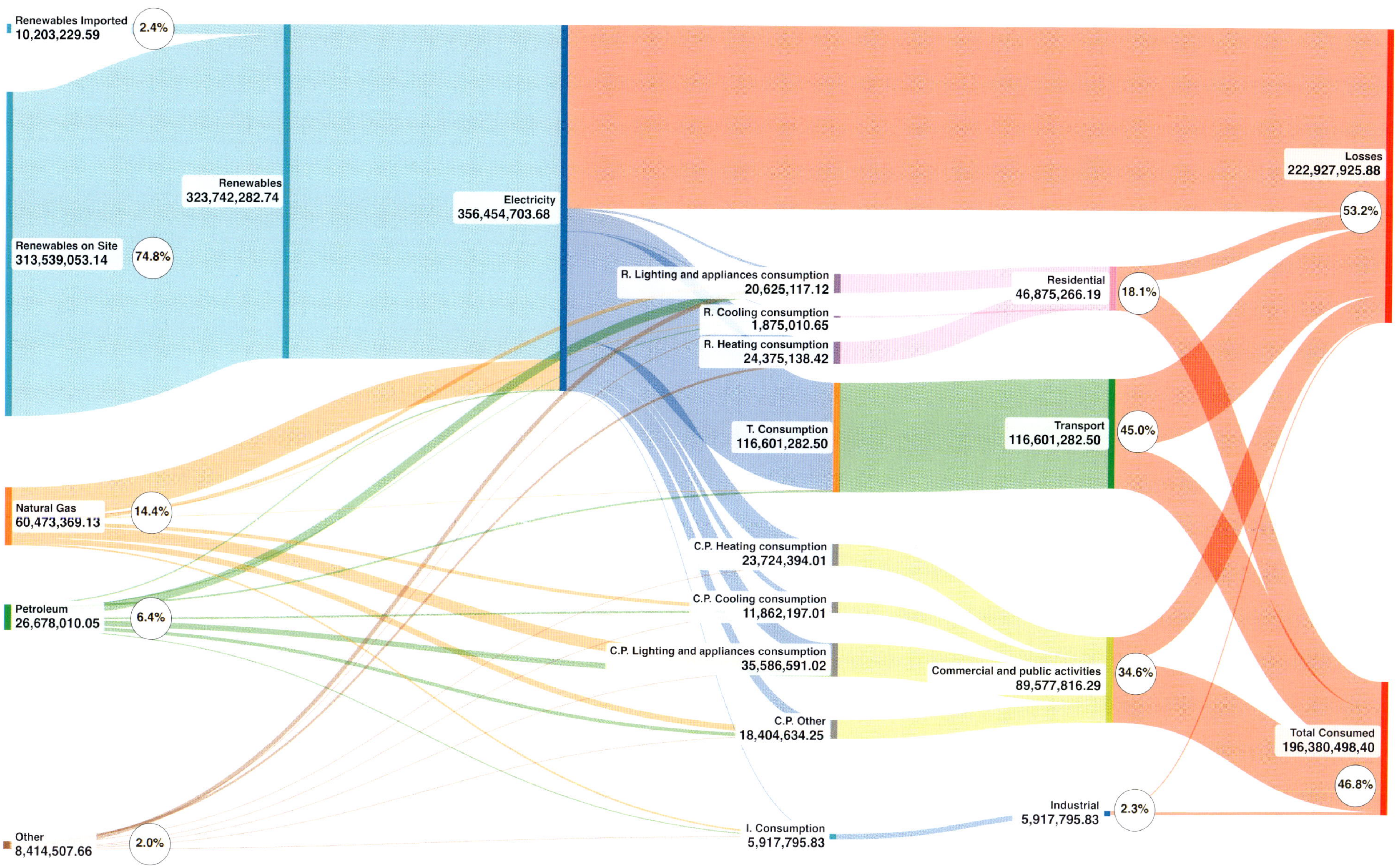

FIG.19: Westwood Energy Efficiency Sankey Diagram.

WESTWOOD MATERIAL MODEL

Usage: **1,838,209 ton/year**
1,838,209,000 kg/year
Population: **7448**

Lifecycle of residential buildings: **61 years**
Lifecycle of commercial and public buildings: **53 years**
Lifecycle of roads: **18 years**
Lifecycle of parking lots: **20 years**
Lifecycle of railway lines: **55 years**

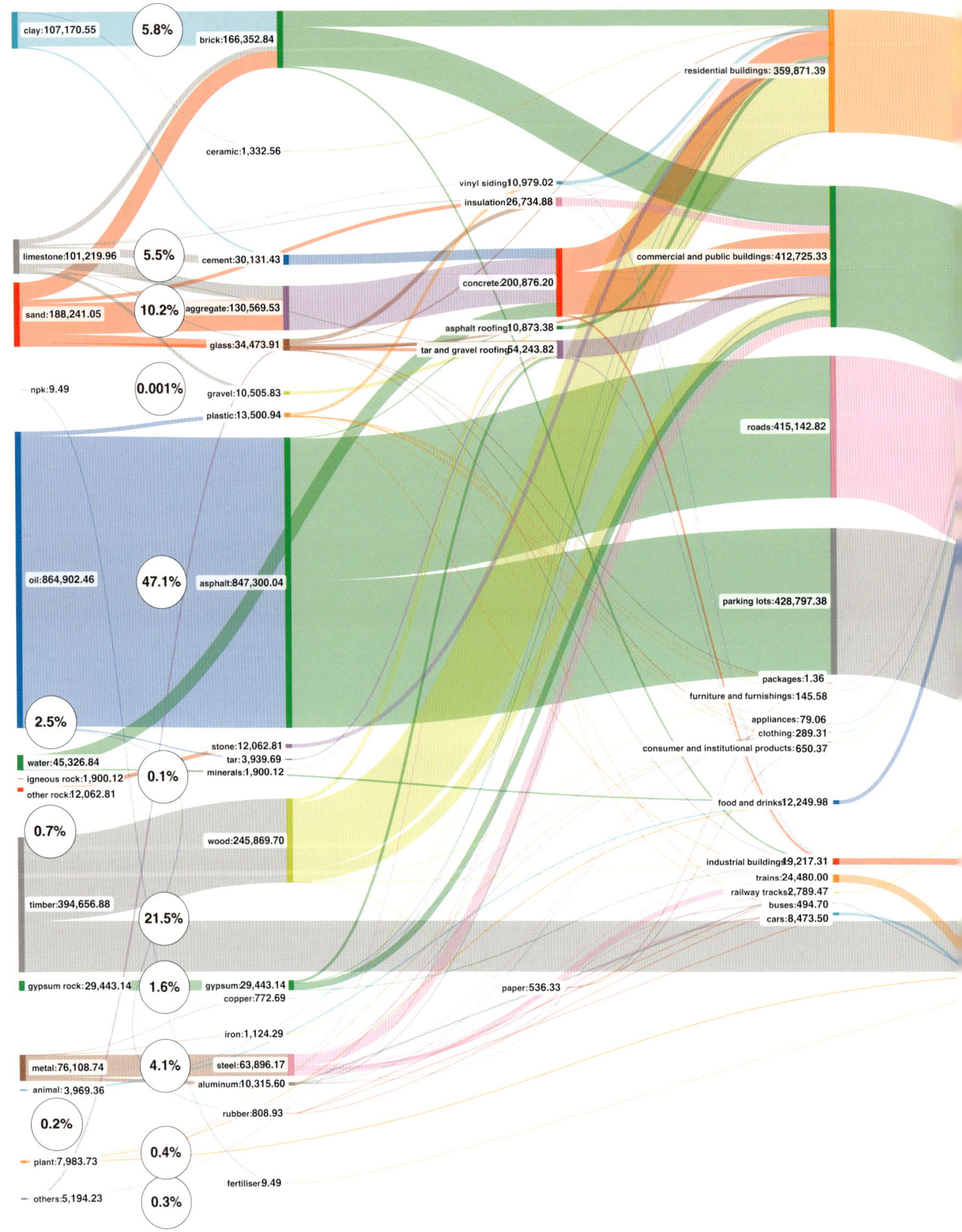

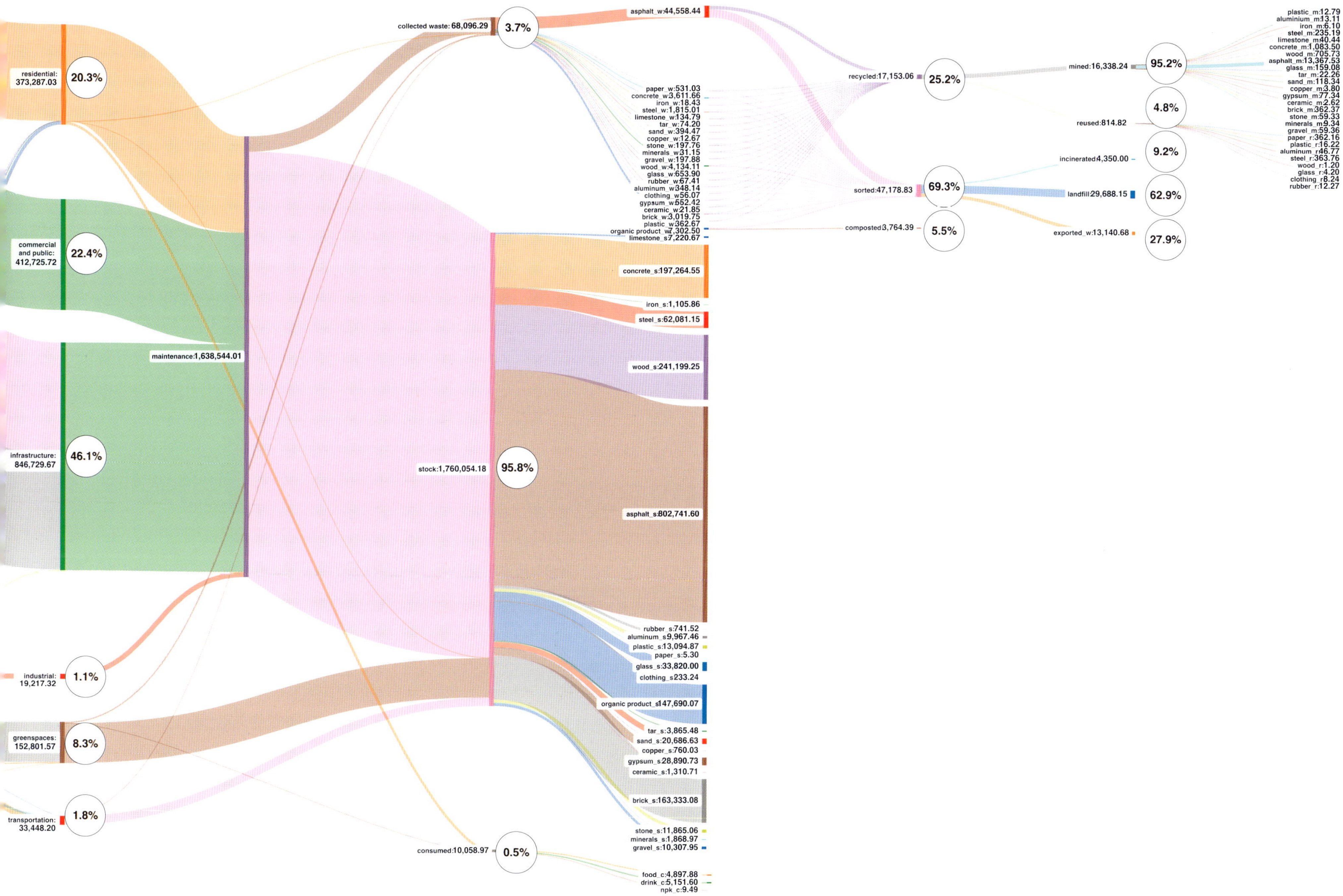

FIG.20: Westwood Material Sankey Diagram.

WESTWOOD						
	Material	**Ton/year (t/year)**	**Embodied energy (MJ/kg)**	**Embodied carbon (kgCO2/kg)**	**Total embodied energy (MJ/year)**	**Total embodied carbon (kgCO2/year)**
1	Asphalt	847300	2.86	0.059	2423278	49990.7
2	tar	3940	44.1	0.39	173754	1536.6
3	plastic	3757	95.4	4.98	358417.8	18709.9
4	glass	34474	15.9	0.9	548136.6	31026.6
5	steel	63896	32	2.7	2044672	172519.2
6	copper	773	44	2.46	34012	1901.6
7	aluminium	10316	154	8.16	1588664	84178.6
8	iron	1124	33	2.5	37092	2810
9	brick	166353	3	0.23	499059	38261.2
10	wood	245870	3	0.3	737609	73760.9
11	minerals	1900	5	0.7	9500	1330
12	rubber	809	110	6	88990	4854
13	ceramic	1333	10	0.66	13330	879.8
14	gypsum	29443	4.4	0.4	129549.2	11777.2
15	stone	12063	0.2	0.1	2412.6	1206.3
16	paper	536	12.6	0.97	6753.6	519.92
17	concrete	200876	0.75	0.1	150657	20087.6
18	vinyl	9744	79.1	2.6	770750.4	25334.4
19	limestone	17861	1	0.1	17861	1786.1
20	sand	21081	0.1	0.005	2108.1	105.4
	Grand Total	**866534**			**9636606**	**542470**
					MJ/kg/year Average 5.76	**kgCO2/kg/year Average 0.32**

FIG.21: Matrix of Materials by Embodied Energy, etc for Westwood.

iii. Organic and Informal Settlements.

The representative site of the 'organic' or 'informal' settlement is the Colonia La Campana in Monterrey, Mexico, and sometimes referred to as 'Cerro de La Campana' because of its settlement on a 'bell-shaped' hill. It is one of the most notable informal neighborhoods in the city located on a steep hill to the south-west of the downtown and with a privileged view of Monterrey as shown in Figures 22 and 23. Like other informal settlements in Monterrey, La Campana is relatively isolated from the municipality, with residents taking on the responsibility for developing and maintaining their neighborhood. It was built through informal, often self-help practices and with limited access to public services, with the exception of water and electricity. Indeed, although 'colonia' means 'neighborhood' in Spanish, it is also the generic term often applied to refer to 'informal settlements' along the U.S.–Mexico border (Richardson and Pisani, 2012). The land area of La Campana is about 23 hectares. With a population of some 6,300 inhabitants across about 1,500 dwelling units, this population is uniformly young with the majority below 29 years of age. It is also mostly a residential community with only 36 or so commercial establishments, although this may change soon (Zapata, 2019; and Trevino, 2020).

Existing public spaces are poorly developed as the narrow spaces between buildings are generally used for vertical circulation up and down the hillside and not for other public activities, as shown in Figures 24 and 25. In keeping with other colonias or informal settlements in Monterrey, inhabitants are relatively low income and the settlement areas attracted immigrants from outside the city. Particularly during the latter part of the 19th century and the early 20th century, Monterrey boomed as a rapidly growing industrial city. Consequently, there was a high demand for cheap labor to both build the city and to man the industrial production. Further, places like the *Barrio Independencia*, formerly the *Barrio San Luisto*, was established immediately adjacent to the construction of the formal city center, in order to house the lower-income working population. Until recently La Campana was a relatively rough neighborhood with substantial disenfranchisement of the younger population. Indeed, this was well documented in the award-winning movie

FIG.22: Aerial View of La Campana in Relationship to Central Monterrey, Mexico.
FIG.23: Oblique View of the Cerro La Campana (Peak) in the Context of Urban Monterrey, Mexico.

FIG.24: Image of Vertical Circulation in La Campana, Monterrey, Mexico.
FIG.25: Image of the Space Between Buildings in La Campana, Monterrey, Mexico.

'Ya no estoy aqui' ('I am no longer here') filmed in 2011 depicting the life of a young gang member with a love for *cumbia*, the local *Kolombian* music and dance, depicted in Figure 26. La Campana was also the subject of sustained community activity with the 'Colosal' painting project that transformed some 300 houses into a multi-colored urban mural project as shown in Figure 27. Also, in 2015 'Barrio Esperanza' was formed to foster community-based improvements, particularly in the sectors of social infrastructure, education, and employment and health (Zapata, 2019). More specifically, within the sector of social infrastructure, efforts are being made to create 'pocket parks' to secure and augment areas especially for children's play, as well as a waste management program focused on recycling. These small and localized efforts have the advantage of being well scaled to local circumstances, stimulating wide community participation and helping to reduce the threat of social displacement (Dovey, 2012).

a. **Water Model**

The defining characteristics of the water stock-flow model depicted in Figure 28 for the 'organic settlement' at La Campana, Monterrey, Mexico, in cubic meters per year are as follows. The percentage distribution of sources are about 62 percent for surface water and 38 percent for groundwater. This is more or less the same as the rest of the city as it is one of the few services provided directly to the informal settlement. The percentage distribution by use is around 83 percent residential, 11 percent commercial-public, and 6 percent on green space. The percentage distribution in output by type is 91 percent to wastewater treatment, 6 percent in leakage, 1 percent in evaporation, and 2 percent in infiltration. These low figures for these last two types are largely because of the absence of open space within the settlement. The total volume for La Campana was 621,000 cubic meters per year. With the use of water-efficient appliances, a reduction of around 42 percent on overall demand could be obtained as shown in Figure 29. Also, as La Campana is fully dependent for water supply and treatment on the surrounding municipality, other potential forms of recycling are outside its jurisdiction as is stormwater harvesting.

b. Energy Model

The defining characteristics of the energy stock-flow model for the 'informal settlement' of La Campana in Monterrey, Mexico, in KWh/year are as follows and are shown on Figure 30. The percentage distribution by source is around 13 percent in renewables, 2 percent nuclear, 85 percent fossil fuels including coal, and 1 percent other. Electricity is really the only other municipal service that is supplied to La Campana along with water. The percentage distribution by use is about 43 percent residential, 12 percent commercial-public, and 45 percent in transportation, reflecting the high dependency on private vehicles within the settlement, essentially to commute outside. The percentage distribution by output is 66 percent lost and 34 percent consumed. For the improved or more efficient energy flow model as shown in Figure 31, the overall amount was potentially reduced by 41 percent. This came about through reduced cooling, heating, and lighting loads by 50, 30, and 70 percent respectively, and the use of ceiling fans, heating devices, and LED lighting. With regard to sources, the potential use of renewables increased to 16.3 percent with solar arrays. Transportation and particularly fossil fuel petroleum use remained high and—same as any energy-saving program—would likely need to be implemented city-wide, such as through an electric bus fleet. The implementation of an energy-reduction program within the informal settlement would likely take the form of self-organization much like the 'pocket park,' colossal painting, and muted waste management programs described earlier.

c. Materials Model

The defining characteristics of the materials stock-flow model in metric tons per year and as shown on Figure 32 for the 'informal settlement' of La Campana in Monterrey, Mexico, are as follows. A full 81.5 percent is drawn from the geosphere reflecting the relatively basic form of development, 2.5 percent is from the biosphere, and 16 percent from the hydrosphere. With regard to land uses, around 67 percent is for residential, 21 percent is in infrastructure, 11 percent in commercial-public, and 1 percent in transportation. With regard to waste management about 93

FIG.26: Characters from the Film 'Ya no estoy aqui' celebrating Cumbia in La Campana.

percent remains in stock, 5 percent is collected waste, and 2 percent is consumed. Of the collected waste, most is not recycled at 92 percent, of which 70 percent is landfilled, 7 percent is recycled, and 1 percent is composted. These statistics also reflect the rather basic and economically depressed circumstances of this informal settlement and that the planned recycling program is yet to get underway. The total material weight was 223,922 metric tons per year.

d. **Embodied Energy and Other Characteristics.**

Figure 33 details the embodied energy in MJ/year and embodied carbon in kg CO_2/year. They average at 2.85 MJ/year and 0.23 kg CO_2/year, respectively, and are the lowest among the four cases, as might be expected.

FIG.27: Results of the 'Colossal' Painting Project in La Campana, Monterrey, Mexico.

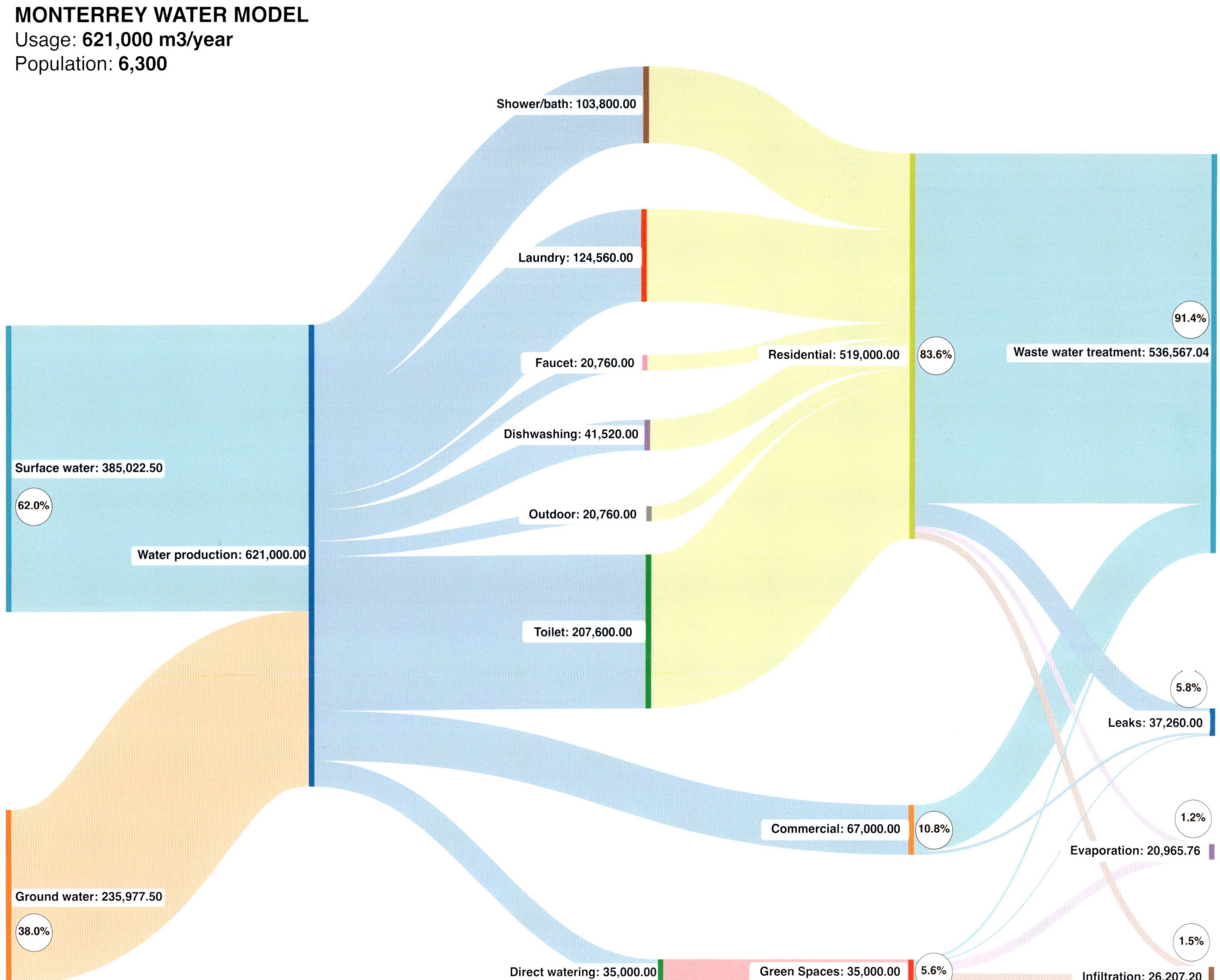

FIG.28: La Campana Water Sankey Diagram.

MONTERREY EFFICIENT WATER MODEL

Usage: **360,614.4 m3/year**
Population: **6,300**

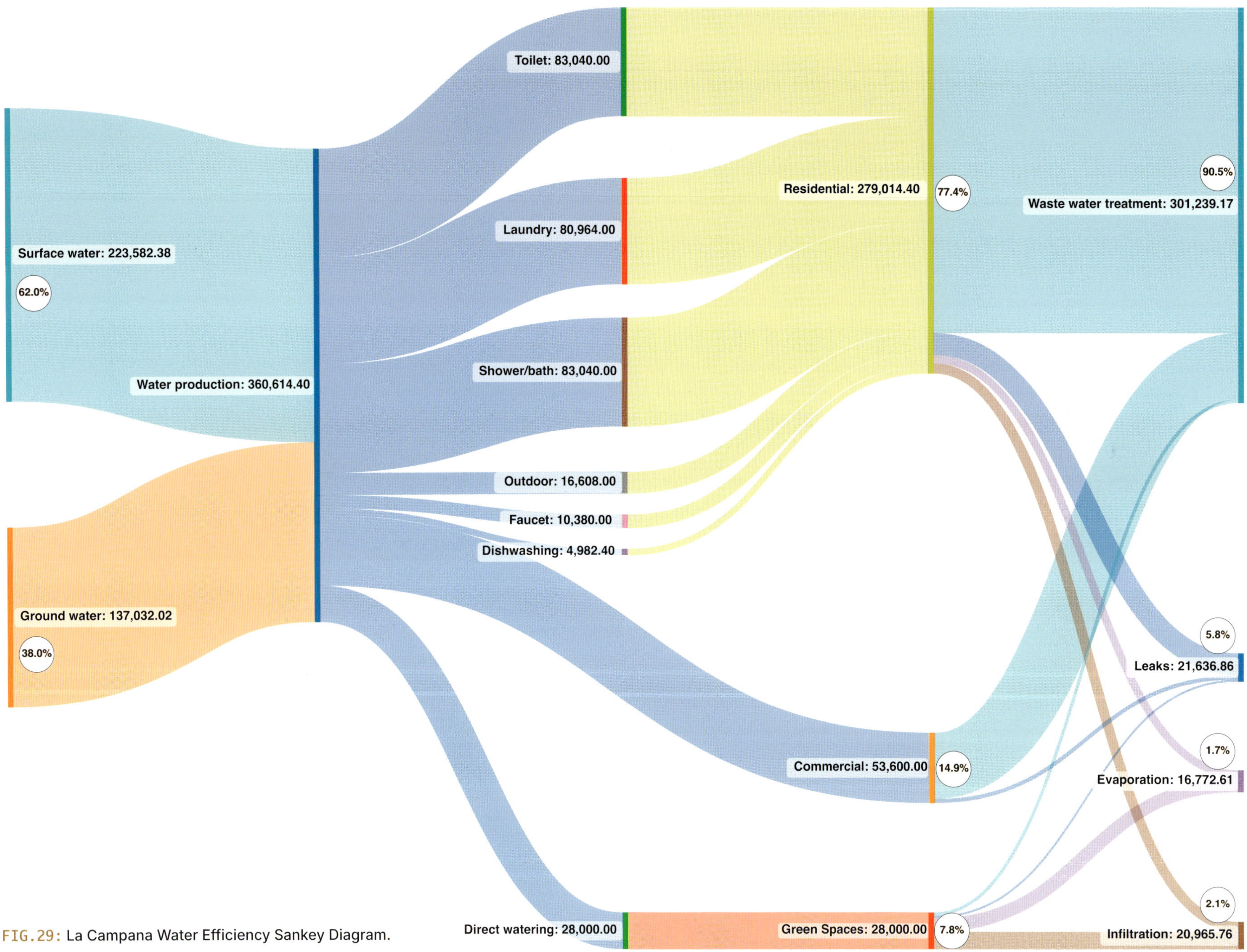

FIG.29: La Campana Water Efficiency Sankey Diagram.

MONTERREY ENERGY MODEL
Usage: **16,742,602.77 kWh/year**
Population: **6,300**

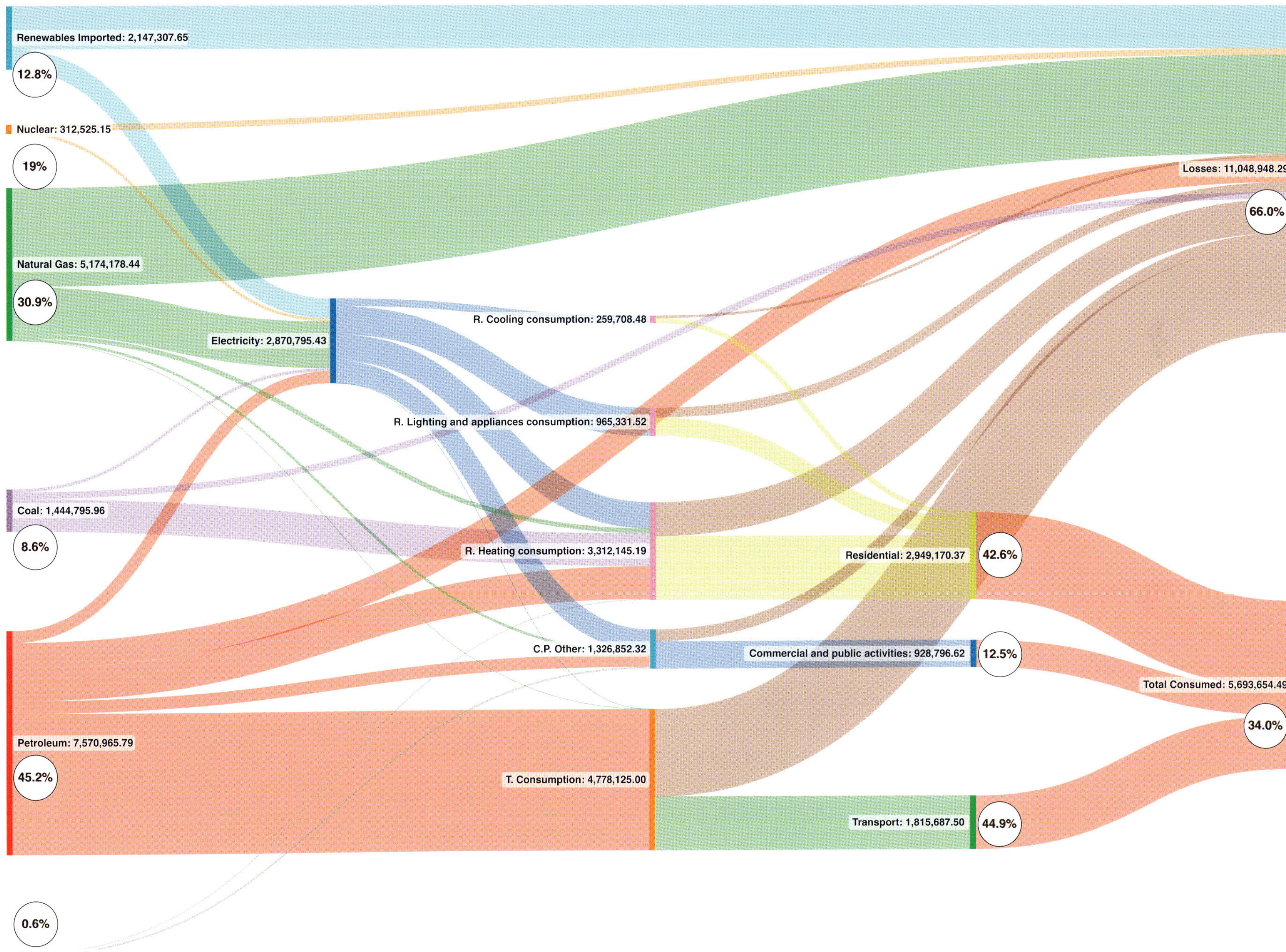

FIG.30: La Campana Energy Sankey Diagram.

FIG.31: La Campana Energy Efficiency Sankey Diagram.

MONTERREY ENERGY EFFICIENT MODEL

Usage: **9,737,161.28 kWh/year**

Population: **6,300**

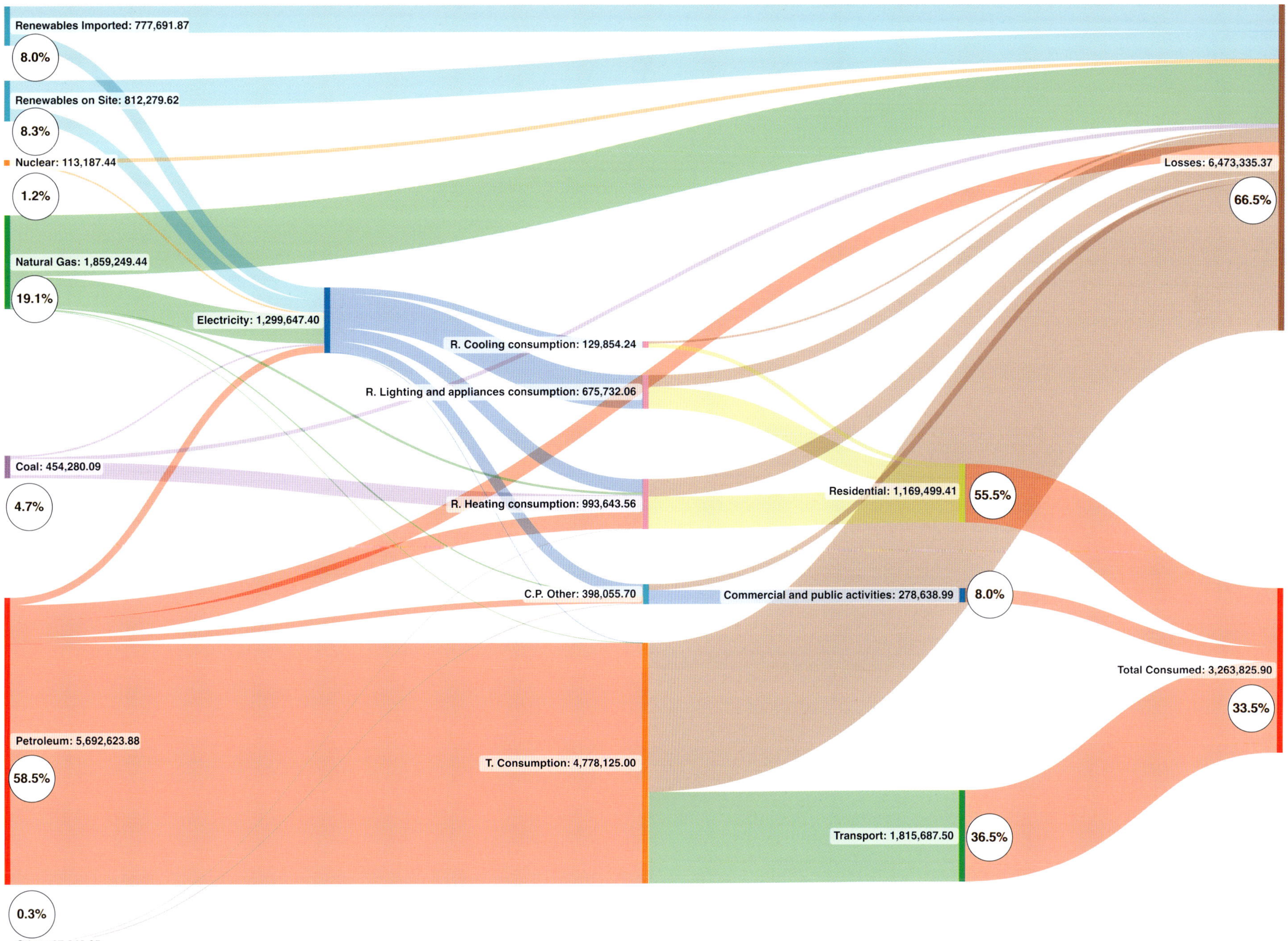

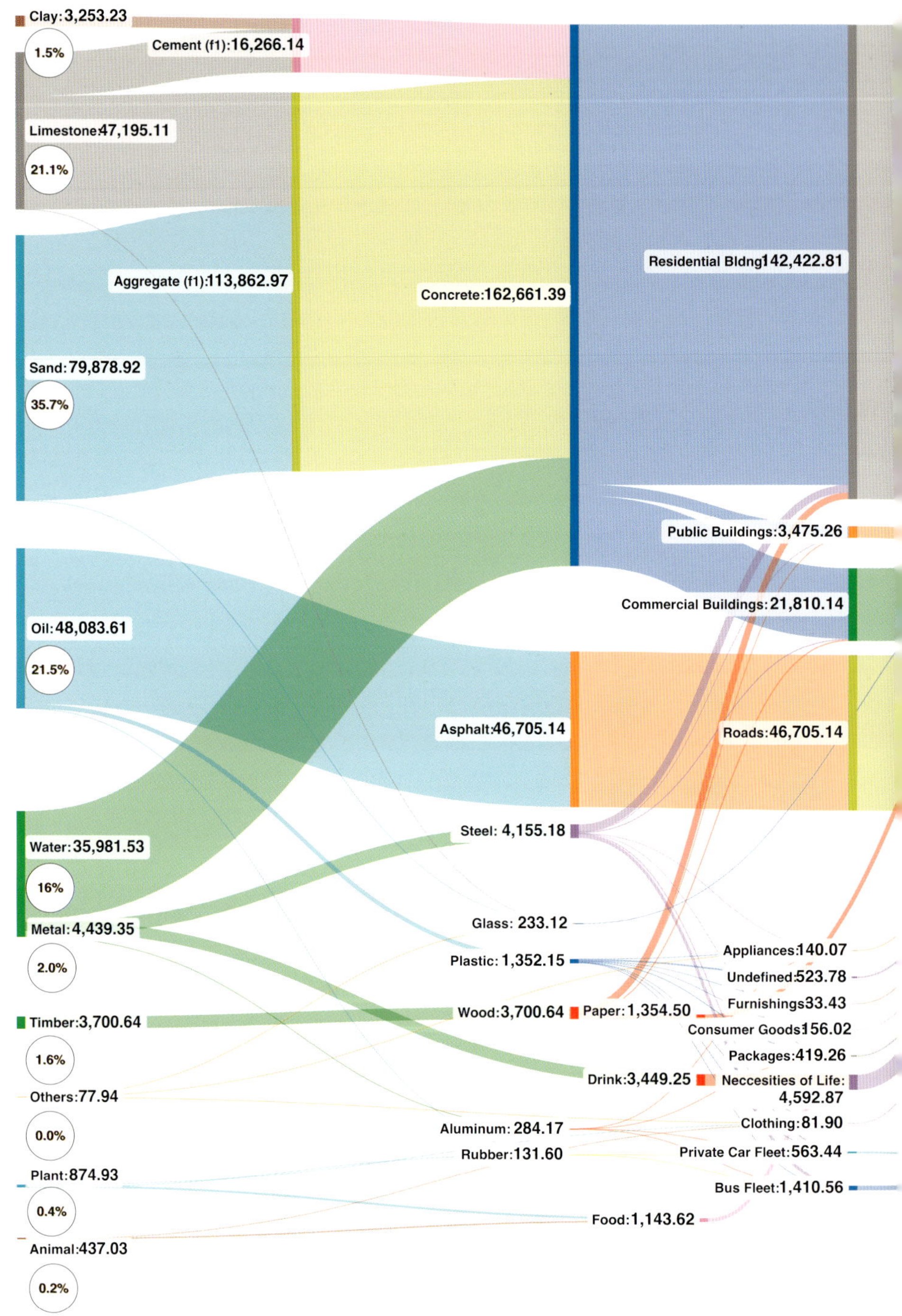

FIG.32: La Campana Material Sankey Diagram.

MONTERREY MATERIAL MODEL

Usage: **223,922.27 ton/year**

223,922,270 kg/year

Population: **6,300**

Lifecycle of buildings: **20 years**

Lifecycle of roads: **20 years**

Lifecycle of cars: **13.33 years**

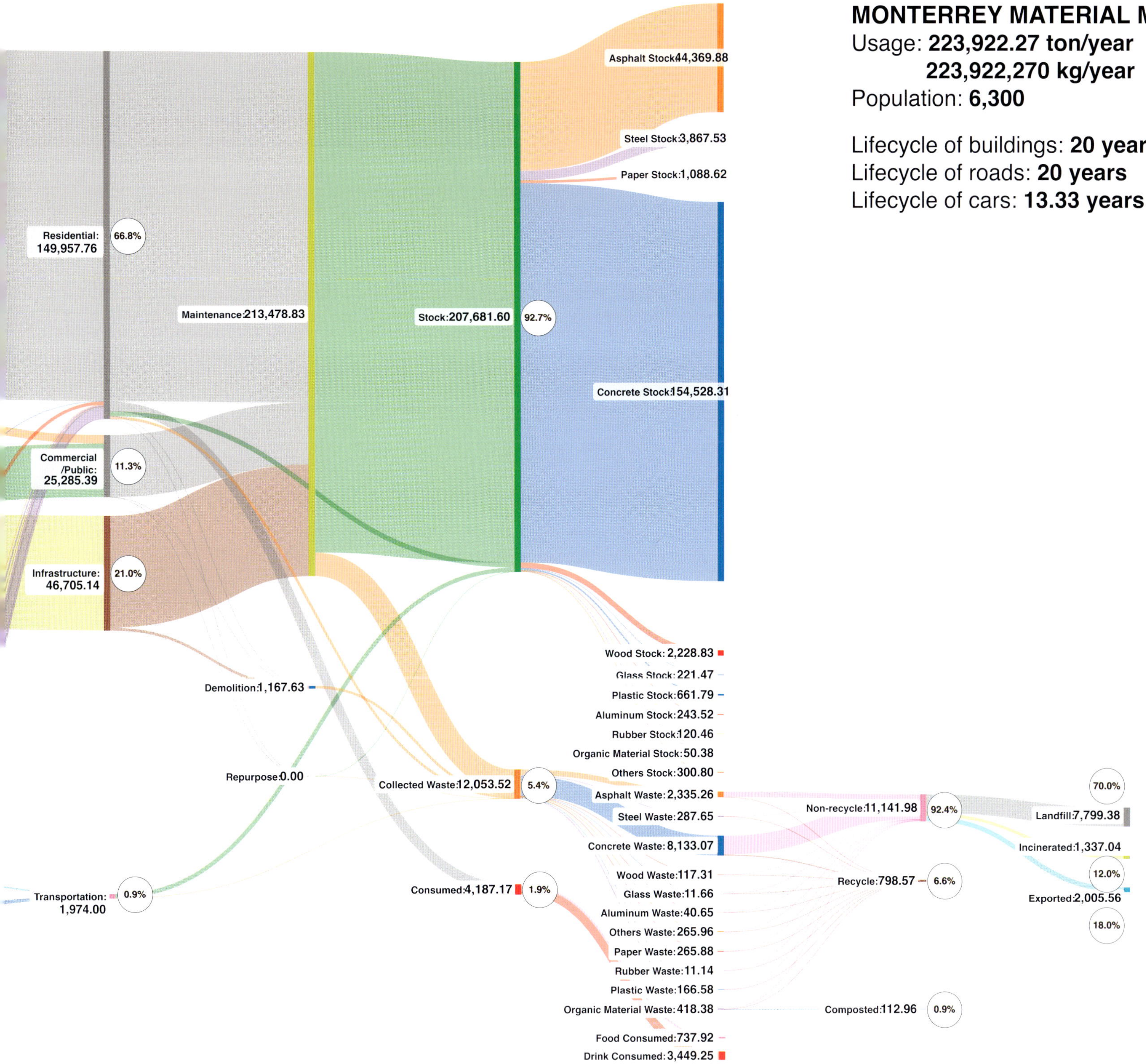

LA CAMPANA - MONTERREY

	Material	Ton/year (t/year)	Embodied energy (MJ/kg)	Embodied carbon (kgCO2/kg)	Total embodied energy (MJ/year)	Total embodied carbon (kgCO2/year)
1	Clay	3253	3	0	9759	748
2	Cement	13013	4	1	52052	9499
3	Concrete	149648	1	0	112236	14965
4	Glass	233	16	1	3730	228
5	Asphalt	46705	3	0	140115	2922
6	Plastic	1352	95	5	128981	6733
7	Steel	4155	32	3	132960	11219
8	Timber	3701	0	0	1110	1110
9	Aluminum	284	154	8	43736	2329
10	Rubber	132	110	6	14520	792
	Grand Total	**224496**			**640621**	**50545**
					MJ/kg/year Average 2.85	**kgCO2/kg/year Average 0.23**

FIG.33: Matrix of Materials by Embodied Energy, etc at La Campana.

iv. Desakota Areas.

The representative site for a desakota area is the village of Chencun in the Shunde District of Foshan City, Guangdong Province, and the Pearl River Delta area of southeastern China. It may also be referred to in China as a 'Township-Village-Enterprise' because of shared characteristics with regard to location and relatively high population and economic growth as shown in Figure 34 (Saich, 2001; and Wu and Sui, 2015). The term *desakota*, from the Indonesian *desa* meaning village and *kota* meaning town, was first coined by Terry McGee in the 1990s (McGee, 1991). For him and others, desakota regions typically happened in Asia and were often located outside the peri-urban zones away from convenient daily commuting and often adjacent to arterial roads. Generally, they were characterized by relatively high population densities and intense agricultural development. They differed from rural areas *per se* because of their more urban-like living conditions. In the Chinese case they occurred primarily during the Reform Period of the 1980s and mostly in coastal areas such as Guangdong Province and they included enterprises usually sponsored by townships and villages. With the modernization and industrialization of China, such as in the Pearl River Delta Region, they also often housed substantial numbers of migrant workers and so-called 'floating populations.' This usually comprised a rentier situation, where the villagers earned rental incomes from the migrants, either working nearby in factories or within village enterprises. As mentioned, the Chencun Village is located in the Shunde District of Foshan, Guangdong Province, as depicted in Figure 35.

Also referred to as *Shuntak* in Cantonese, this district is a county-level city with an area of 806 square kilometers and a population of 2.5 million inhabitants, renowned for various forms of agriculture from the late Ming through the Qing Dynasties; it was a center of sericulture and silk production. Indeed, the Chencun Village remains an agriculturally centered production site, with its footprint covering 60 percent of the site, now for fresh flowers and trees as depicted in Figure 36. In fact, the value of agricultural production for China, judging from 2016 data, is the highest by nation in the world, as shown in Figure 40 (Food and Agriculture Organization, 2018). Also as shown in Figure 36,

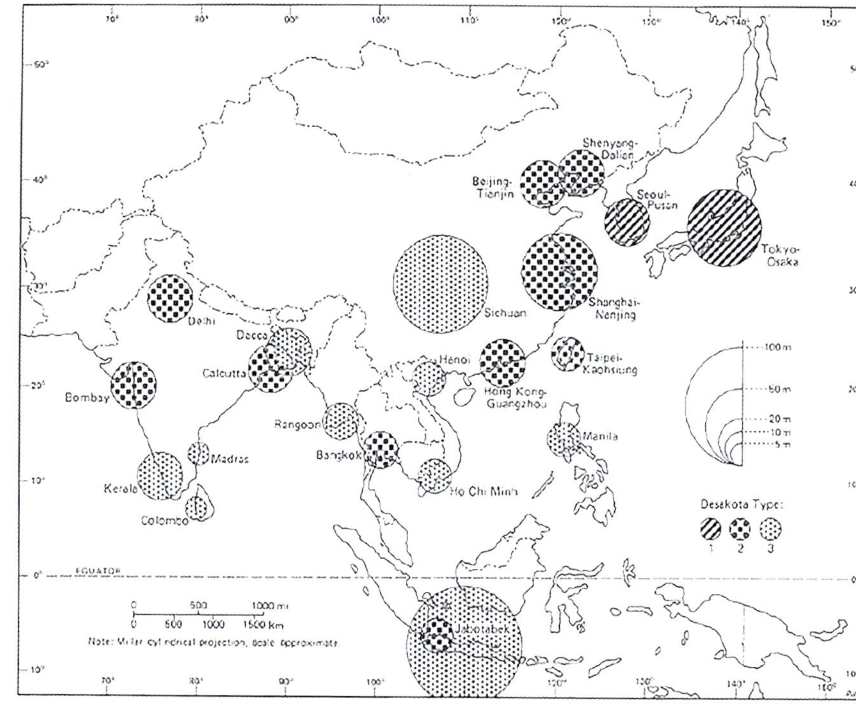

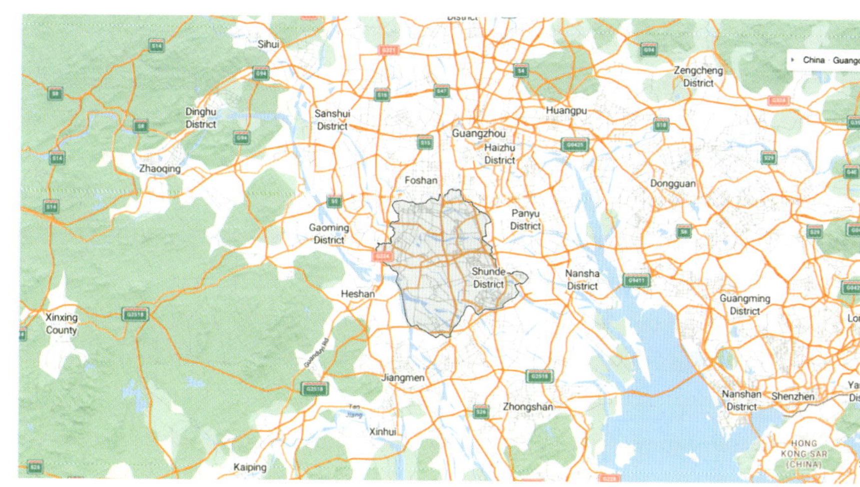

FIG.34: Three Types of Desakota Patterns in Asia.
FIG.35: Shunde District in the Context of Guangzhou and Guangdong Province, China.

Chencun, Shunde District
Foshan City, Guangdong Province, PRC

FIG.36: The Layout of Chencun Village, Shunde District in Guangdong Province, China.
FIG.37: Images of Residential Buildings in Chencun Village, China.
FIG.38: Images of Commercial Uses in Chencun Village, China.

the village, covering an area of 131 hectares, takes on a dispersed but contiguous overall planar form, mainly of residences, often associated with district-level roadways. Adjacent spaces are given over to plants and tree growing, with a large flower market located on Weixin Lou, a major arterial that cuts through the village from west to east. The population is comprised of some 5,000 village residents and a 'floating population' of another 1,000 or so persons for a total of some 6,055 inhabitants across 1,781 dwelling units. For the most part they are housed in three- and four-story housing units with the ground floors occupied by a garage or a commercial enterprise along one of the main roads, as indicated in Figure 37, or without any non-residential use. The 'floating population' usually occupies units belonging to permanent residents either together with them or as renters of separate units. Typical housing units are shown in Figure 37. Generally, the resident community is well-off economically and also incorporates some small factories making hardware, as well as hosting a primary school and numerous commercial outlets depicted in Figures 37 and 38. Nevertheless, in keeping with its desakota designation, the primary means of production is agriculture as shown in Figure 39.

a. **Water Model.**

The defining characteristics of the water stock-flow model for the 'desakota area' at Chencun in China, in cubic meters per year, are as follows and are depicted in Figure 41. The distribution by source is 46 percent surface and 54 percent stormwater, reflecting the agriculture associated with a 'desakota area.' The distribution by use is 21 percent residential, 3 percent commercial public, 2 percent industrial, 1 percent public open space, and 73 percent in agriculture. The distribution of output by type is about 28 percent from wastewater treatment, 9 percent leaked, 28 percent evaporated, and 35 percent from infiltration, again the last two statistics reflecting the agricultural aspect of 'desakota areas.' Also, again the reduction in water use through graywater recycling from urban uses is relatively minimal proportionally, at about 12 percent as shown in Figure 42.

FIG.39: Images of Agricultural Production in Chencun Village, China.
FIG.40: Worldwide Value of Agricultural Production, 2016.

Value of agricultural production, 2016

Gross production value of the agricultural sector, measured in current US$.

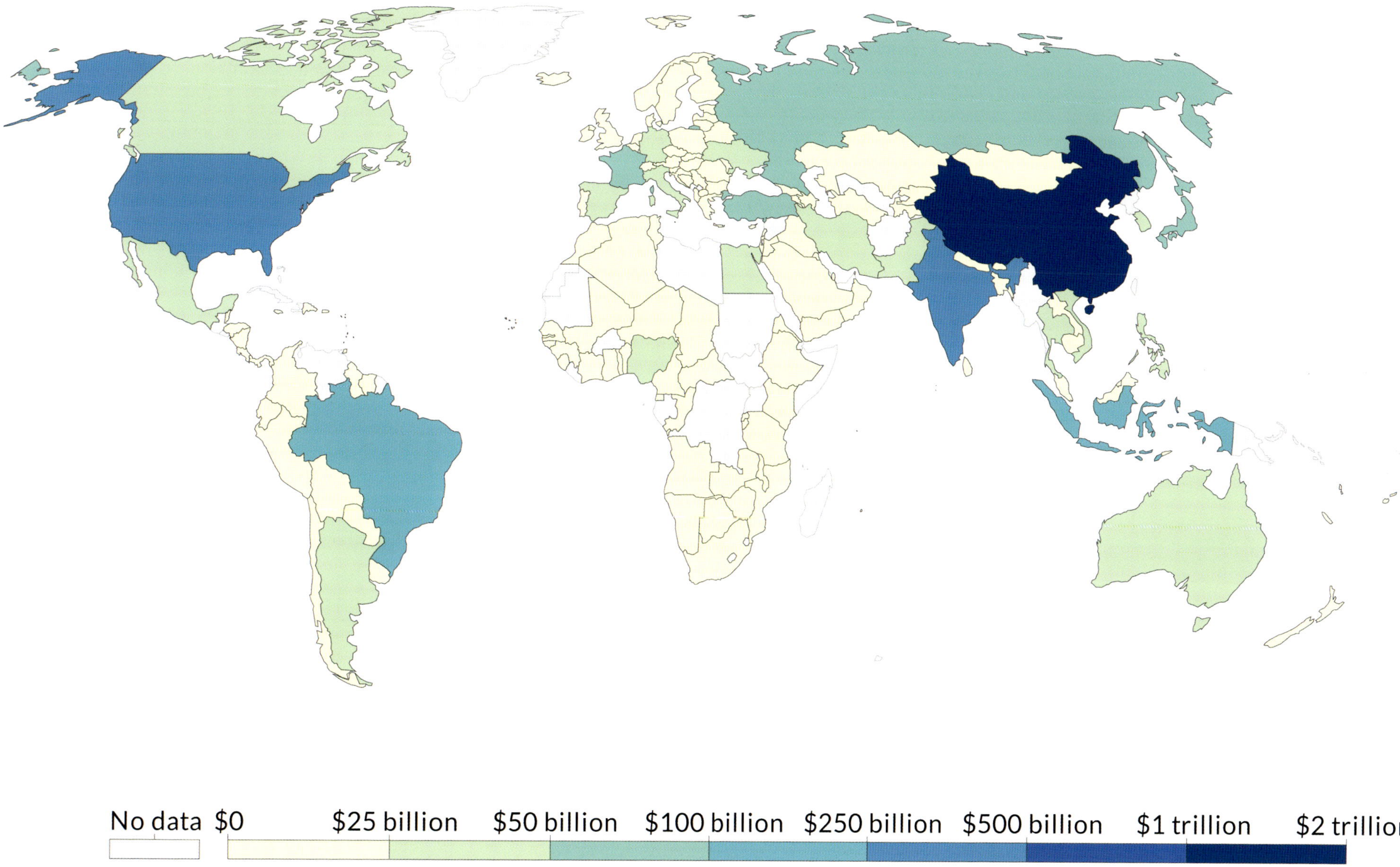

CHENCUN WATER MODEL

Usage: **2,008,220 m3/year**
Population: **6055**

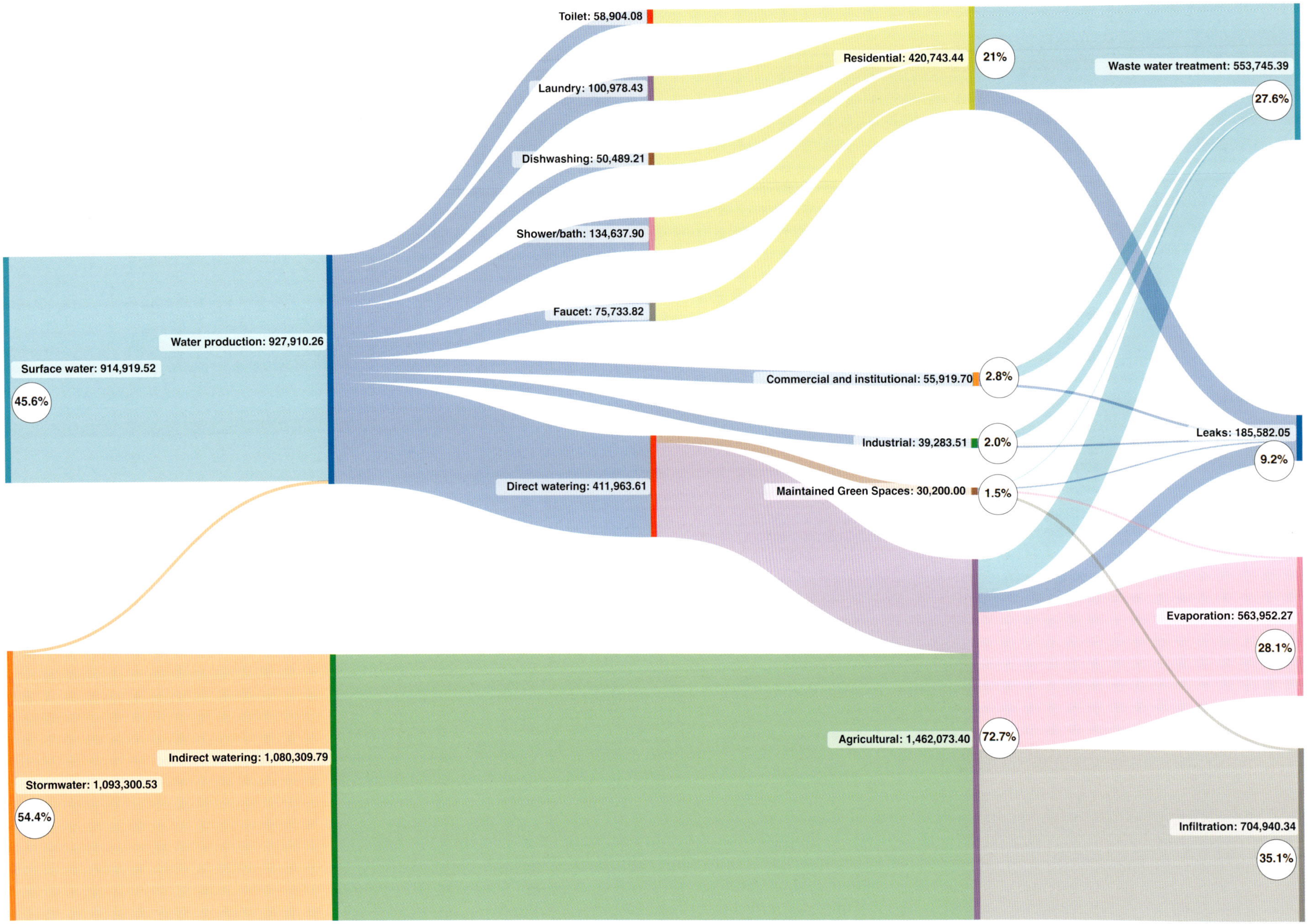

FIG.41: Chencun Water Sankey Diagram

CHENCUN EFFICIENT WATER MODEL

Usage: **1,737,203 m3/year**

Population: **6055**

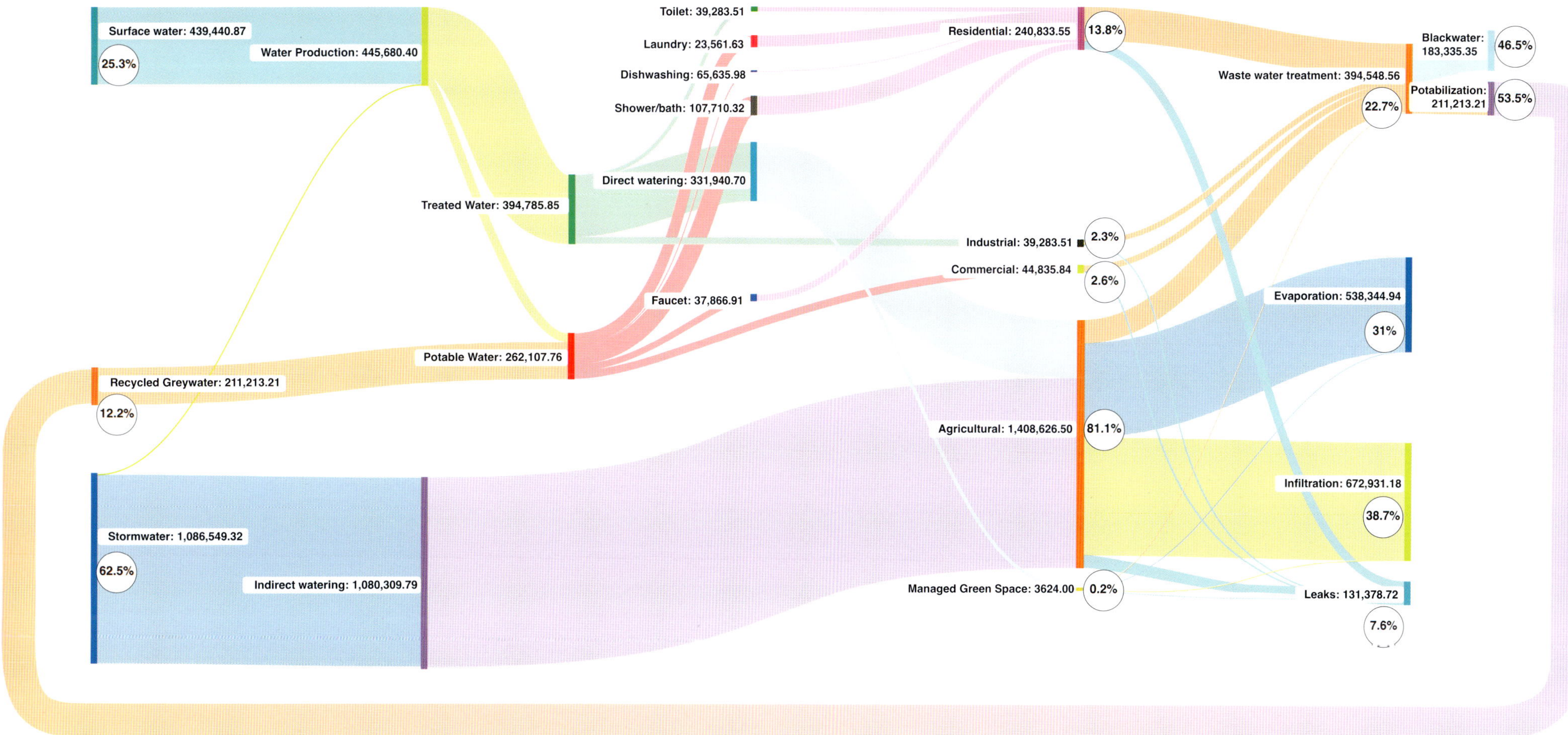

FIG.42: Chencun Water Efficiency Sankey Diagram

CHENCUN ENERGY MODEL

Usage: **265,330,593 kWh/year**

Population: **6055**

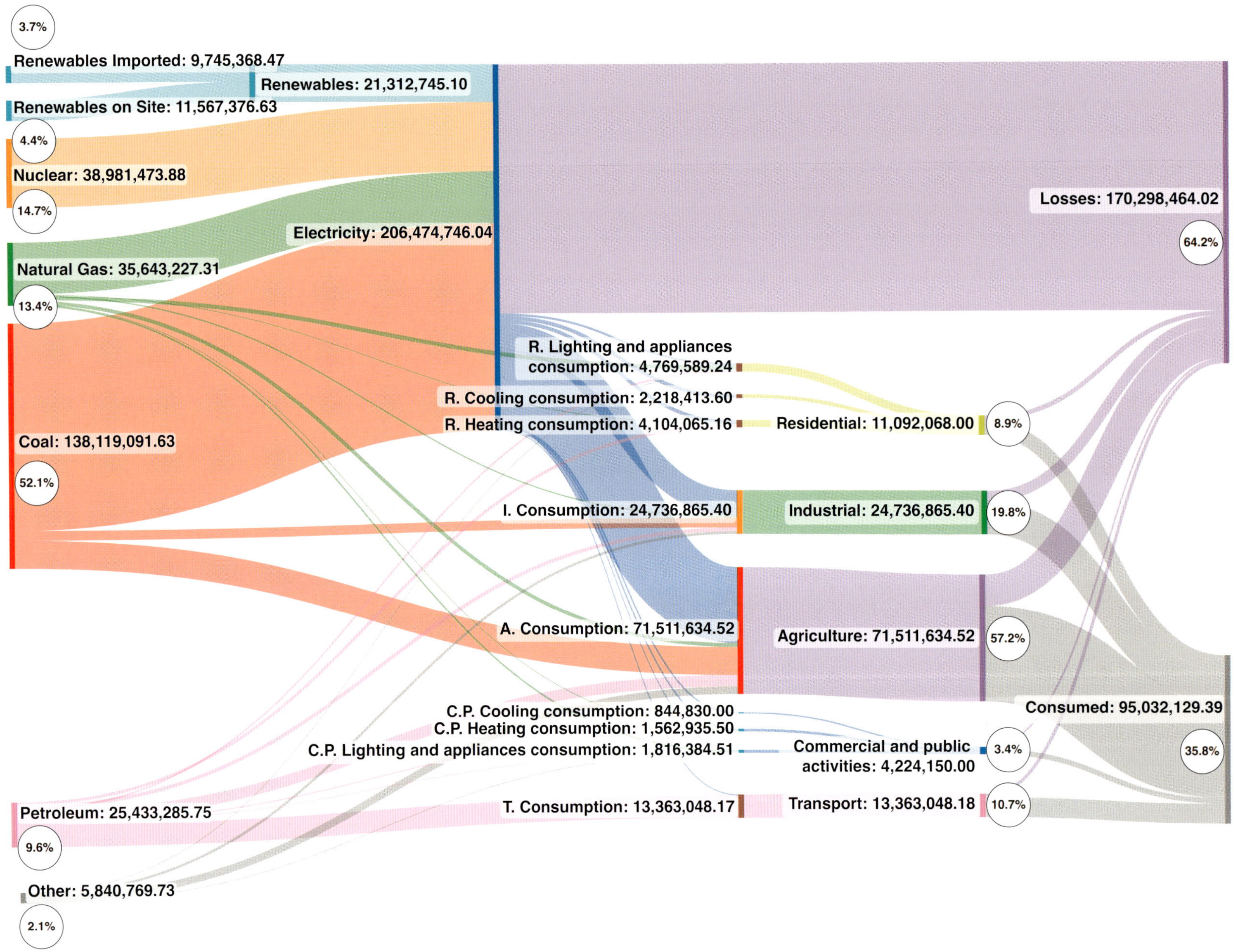

FIG.43: Chencun Energy Sankey Diagram

CHENCUN ENERGY EFFICIENT MODEL

Usage: **265,330,593 kWh/year**
Population: **6055**

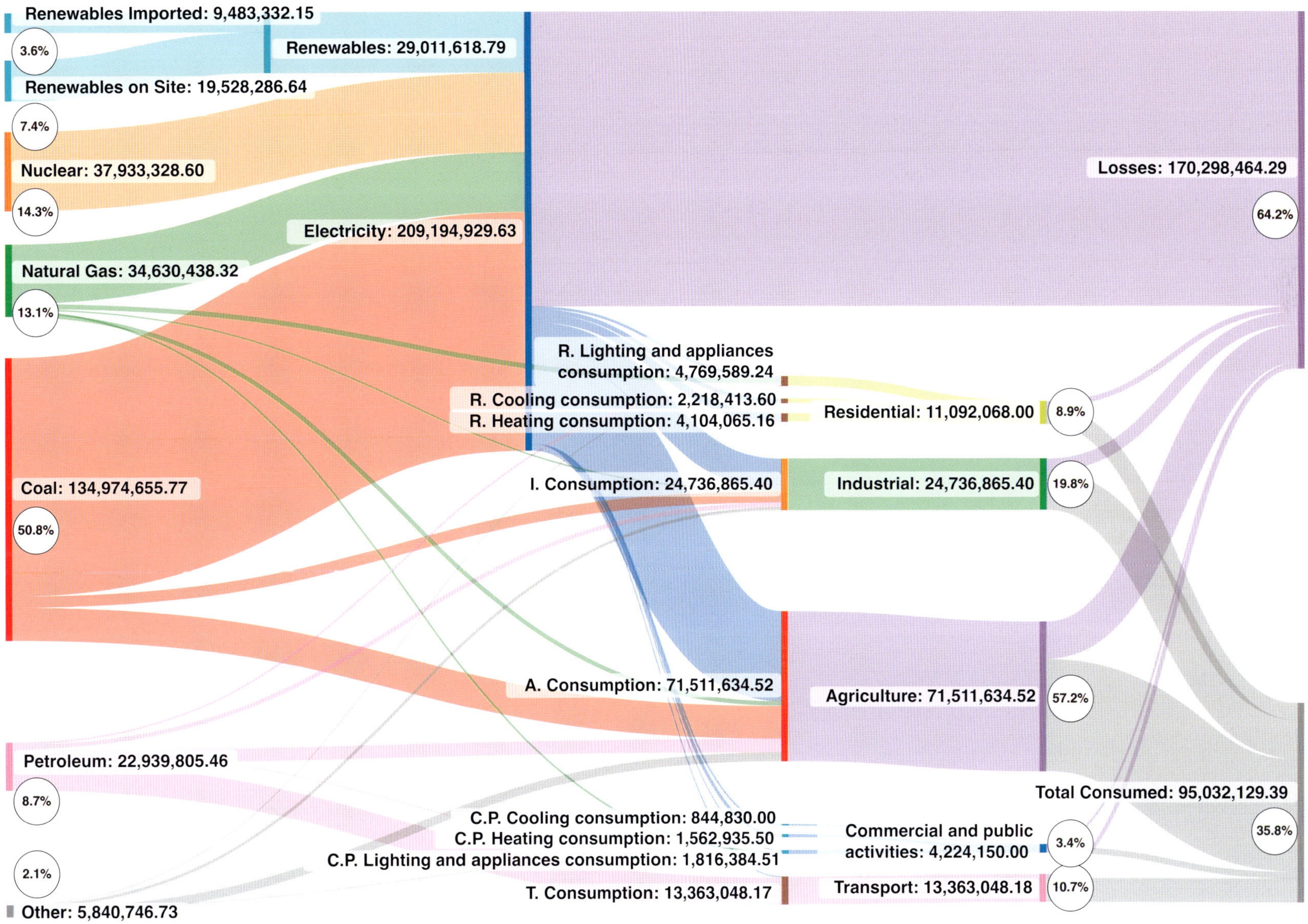

FIG. 44: Chencun Energy Efficiency Sankey Diagram

CHENCUN MATERIAL MODEL

Usage: **705,961 ton/year**
705,961,000 kg/year
Population: **6055**

Lifecycle of buildings: **35 years**
Lifecycle of hardscape: **20 years**
Lifecycle of roads: **15 years**
Lifecycle of parking lots: **50 years**
Lifecycle of railway lines: **50 years**

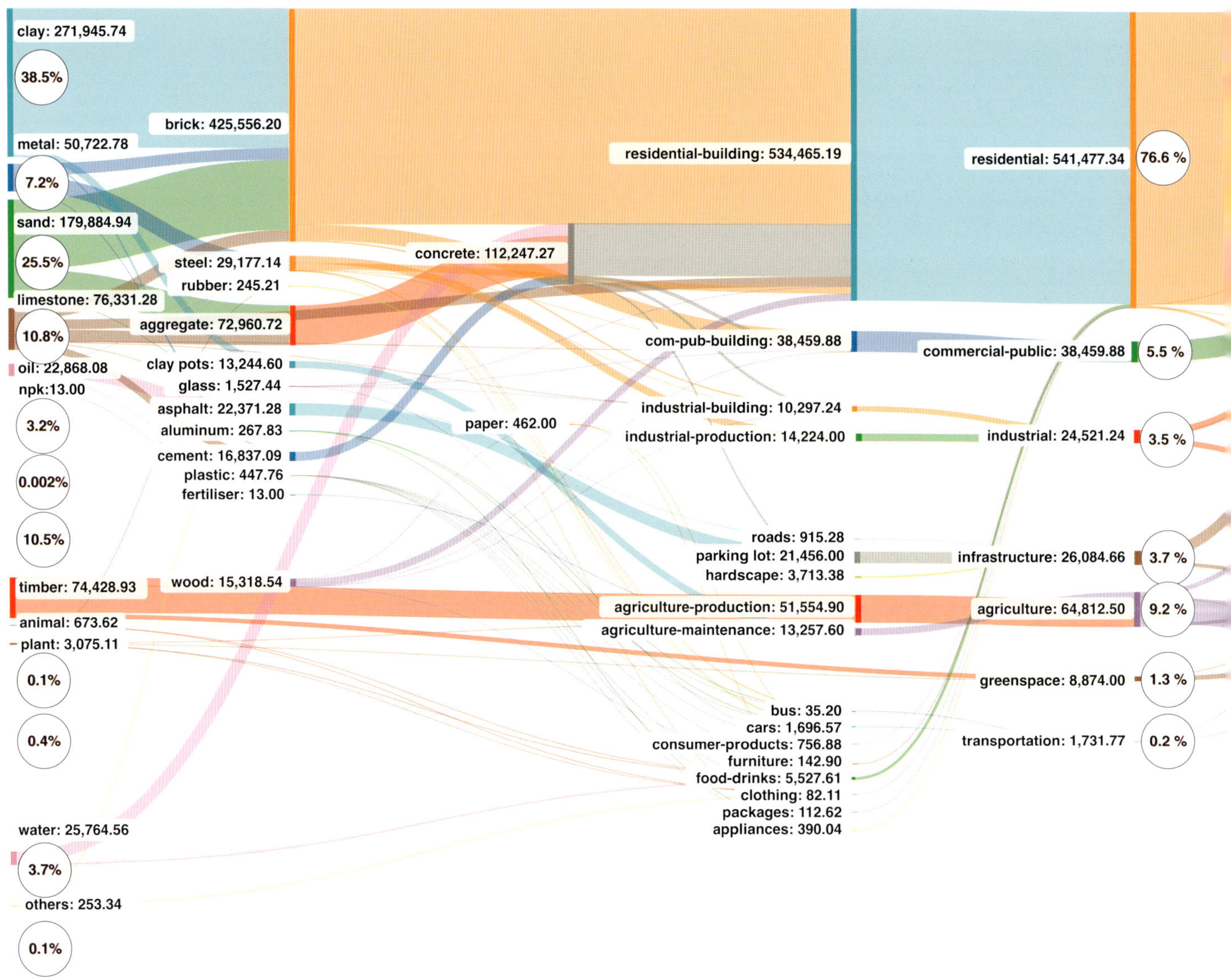

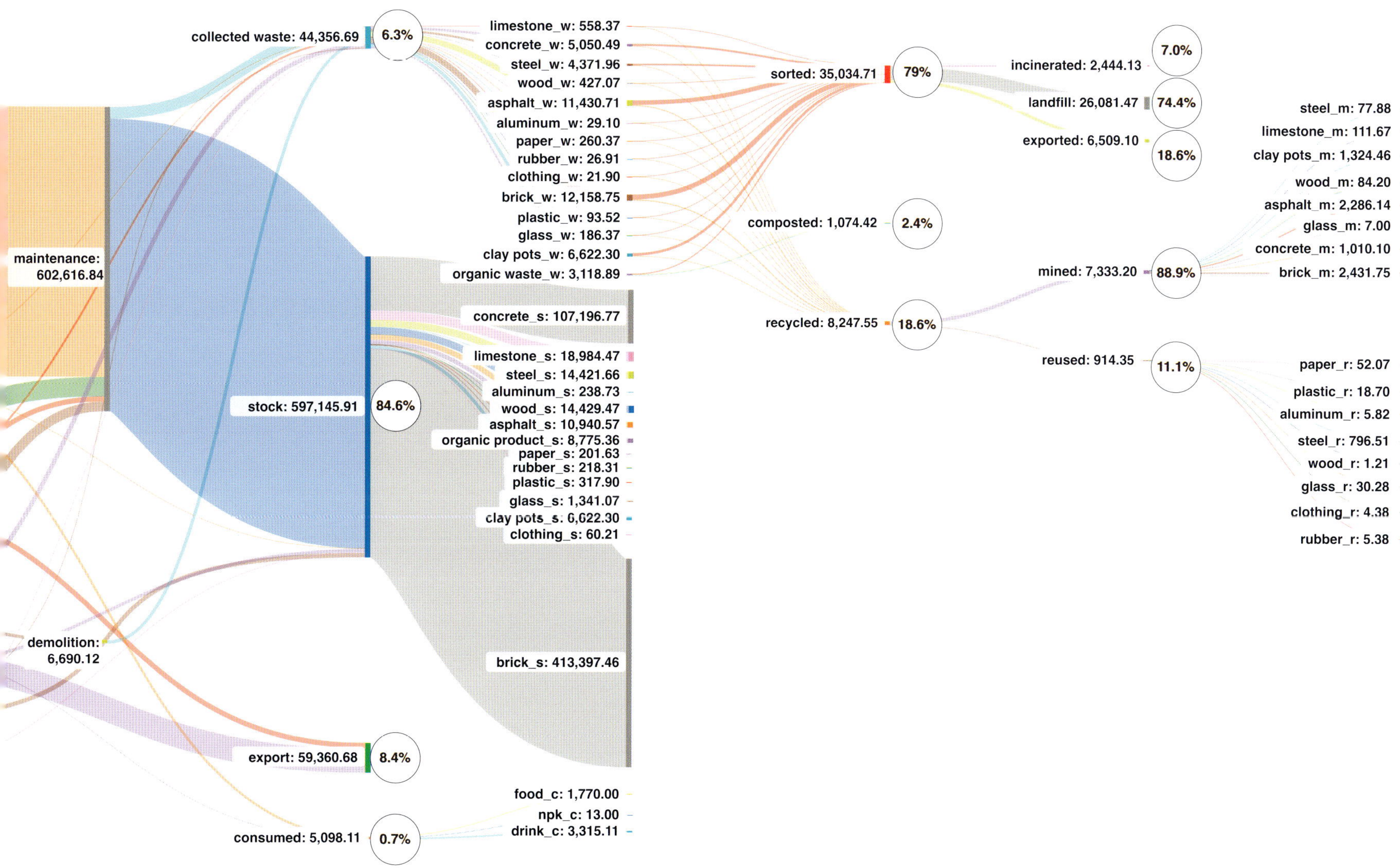

FIG.45: Chencun Material Sankey Diagram

b. Energy Model

The defining characteristics of the energy stock-flow model for the 'desakota area' at Chencun in China are as follows and are shown in Figure 43. The distribution by source is around 8 percent in renewables, 15 percent from nuclear, 75 percent in fossil fuels with 52 percent from coal (not uncommon in China), and 2 percent from other (probably biomass). The distribution by land use is 9 percent residential, 20 percent industrial, 57 percent agricultural, 3 percent commercial-public, and 11 percent in transportation, again reflecting the 'desakota area' character of settlement. The distribution of output is 64 percent lost and 36 percent consumed. For the improved or efficient energy model as depicted in Figure 44, there was minor diminution of use due to on-site renewable sources for heating and cooling.

c. Materials Model

The defining characteristics of the material stock-flow model in tons per year, for the 'desakota area' at Chencun, are as follows and are shown in Figure 45. A full 85 percent was drawn from the geosphere, 11 percent from the biosphere, and 4 percent from the hydrosphere. With regard to land use, most, at 77 percent, was in residential; 6 percent was in commercial-public; 3 percent in industrial; 4 percent in infrastructure; 11 percent in agriculture and green space, seemingly low but in weight of material and not area of use; and with 1 percent in transportation. With regard to waste management, 85 percent remained in stock, 8 percent was exported, 1 percent consumed, and 6 percent collected as waste. Of the collected waste, 79 percent was sorted, 19 percent recycled, and 2 percent composted. Of the sorted waste 74 percent was landfilled, 19 percent exported, and 7 percent was incinerated. Of the recycled waste, 89 percent was mined and 11 percent reused, again reflecting the character of 'desakota areas.'

d. Embodied Energy and Other Characteristics.

The attached figure details the embodied energy in MJ/year and the embodied carbon on kg CO2/year for Chencun in China. The average is also comparatively low at 3.98 MJ/year in embodied energy and 0.32 kg CO2/year in embodied carbon and are shown in Figure 46.

CHENCUN						
	Material	**Ton/year (t/year)**	**Embodied energy (MJ/kg)**	**Embodied carbon (kgCO2/kg)**	**Total embodied energy (MJ/year)**	**Total embodied carbon (kgCO2/year)**
1	Steel	29177	32	2.7	933669	78778
2	Aluminum	268	154	8.16	41246	2186
3	Wood	65554	3	0.3	19666	15318
4	Plastic	448	95.4	4.98	42716	2230
5	Asphalt	22371	2.86	0.059	63982	1320
6	Brick	425556	3	0.23	1276669	97878
7	Rubber	245	110	6	26950	1470
8	Glass	1527	15.9	0.9	24286	1375
9	Concrete	112247	0.75	0.1	84185	11225
10	Limestone	19543	1	0.1	19543	1954
11	Clay	13245	3	0.23	39734	3046
	Grand Total	**690181**			**2749642**	**221128**
					MJ/kg/year Average 3.98	**kgCO2/kg/year Average 0.32**

FIG.46: Matrix of Materials by Embodied Energy, etc at Chencun

v. Comparative Analyses.

Among the three Sankey stock-flow models for each of the four cases, perhaps the most telling are the material flow models. As already discussed, the population within each is more or less within the same range of 6,000 to 7,500 inhabitants. They are all 'local' in this regard. However, they vary in density and distribution of uses along with accumulated material weights associated with those uses, largely according to the *a priori* definition given to them as representative of different domains of settlement co-existence. The Paris ZAC, for instance, is dominated by the facilities of 'compact cities' and daily life in residences and commercial-public uses, accounting for 94 percent of material weight. Westwood, on the periphery of Boston and as a 'dispersed city' form is dominated by the infrastructure of roads and parking lots, accounting for a massive 45 to 46 percent of total development in material weight. La Campana in Monterrey, Mexico, as an informal organic settlement in relatively precarious economic circumstances, is primarily comprised of residences at 66.8 percent by weight and little else, except for some commercial uses largely within the scope of the residential development. Chencun in China, by contrast with the other cases, is a desakota area and dominated by its agricultural business at 11 percent by weight and its adjacent residential development at 77 percent by weight that also houses a rent-paying 'floating population.' Looking across all four cases, as one might expect, La Campana in Monterrey is low in constructed weight at only 35.5 metric tons per population and only 7 percent of the total weight of the four cases. By contrast, relatively affluent Westwood in the U.S. is probably profligate in consumption and with large houses and extensive infrastructure at some 246.8 metric tons per person and 48 percent of the total weight of the four cases. The ZAC in Paris and Chencun in China occupy a middle ground, so to speak, at 116.8 metric tons per person, or 23 percent of the total and 116.6 metric tons per person, or 22 percent of the total, as shown in Figure 47.

After applying standard multipliers for embodied energy in MJ/kg of material and embodied carbon in kg $CO2$/kg of material, several observations can be made. First and moving towards summary comments in Chapter 6 dealing with 'emergent issues,'

FIG.47: A Summary of Stock-Flow Characteristics by Case.
FIG.48: The Relative Distribution of Materials by Embodied Energy Coefficient.

	Tns/yr/pop	AveMgJ/kg/yr	AveKCO2/kg/yr	% of Total Weight	% of Total Weight < 5MJ/K
Paris ZAC	116.8	3.2	0.28	23	39
Westwood	246.8	5.8	0.32	48	25
La Campana	35.5	2.8	0.23	7	50
Chencun	116.6	4.0	0.32	22	38

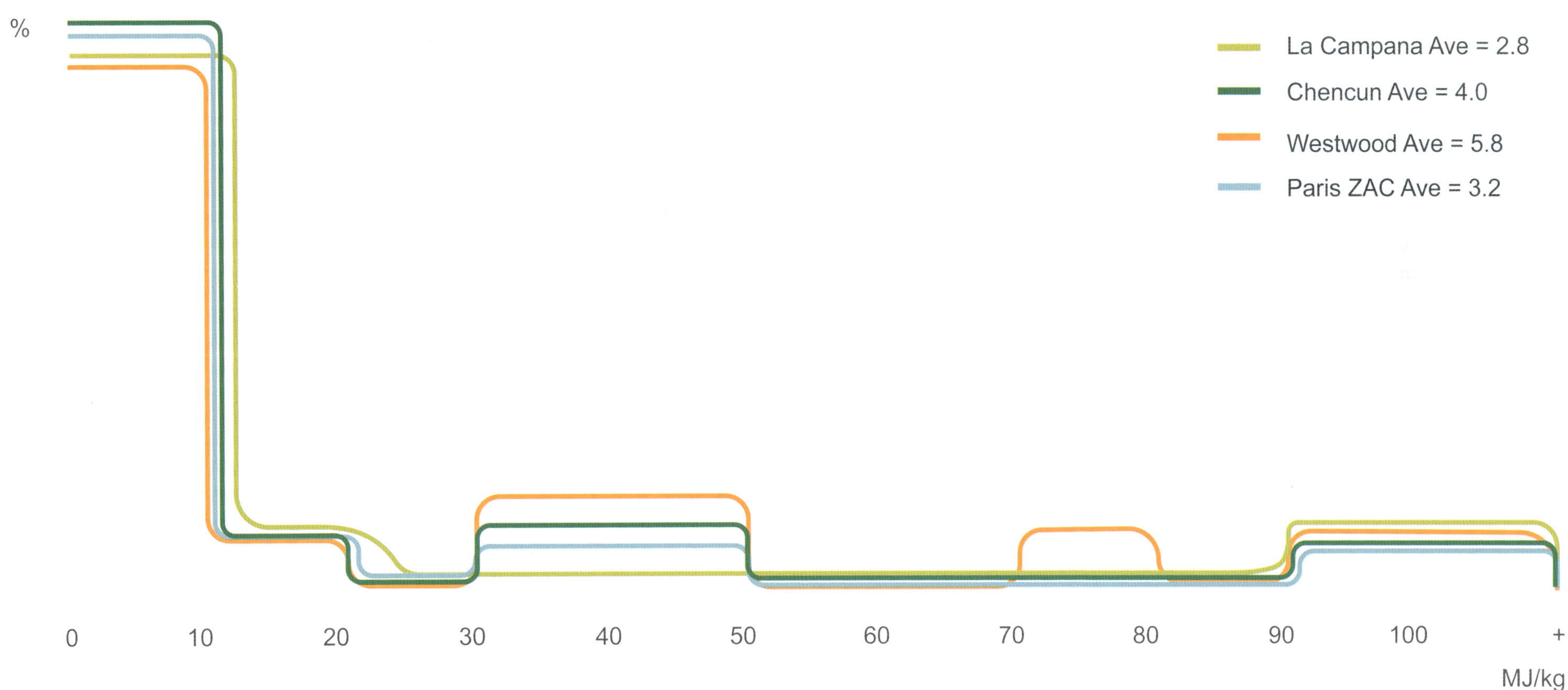

	Buildings	Roads	Transport
Paris ZAC	91	15	9
Westwood	54-61	18	12
La Campana	20	20	13.3
Chencun	35	15	9

FIG. 49: A Summary of Obsolescence Rates by Component and by Case.

application of these multipliers clearly reflects the amount of energy used and carbon emitted in the manufacture of certain materials. In the case of the Paris ZAC, for instance, concrete comprises 76.8 percent of the total material weight as it also might elsewhere in the modern world but contributes only 35.5 percent of the total embodied energy. By contrast, in the same locale aluminum is sparingly used at less than 1 percent of total material weight but amounts to 4.5 percent of the embodied energy. Second, probably a better measure of the environmental performance of each case is via the computed average amount of embodied energy and carbon. Again, as shown in Figure 47, the average for Westwood is the highest at 5.8 MJ/year, reflecting the high levels of consumption and uses of materials in suburban American circumstances and the amount of support, as it were, required to sustain such development. La Campana, at the other more precarious end of the scale, is low at only 2.8 MJ/year on average with the Paris ZAC with all its environmentally friendly inclusions at a respectable 3.2 MJ/year and with Chencun at 4.0 MJ/year. Indeed, this result for the Paris ZAC is considerably less by around 30 percent from the amount of 4.1 for the larger area in Batignolles.

To put these figures further into perspective, looking across a data set covering both kinds of multipliers by specific materials and ranging from, say, 1.0 for aggregate up to as high as 154 MJ/kg for aluminum in embodied energy, the highest average at Westwood at 5.8 MJ/kg/year is in the range of material performances of, say, certain timber formats at 2.5 to 8.5 MJ/kg and the figure for La Campana comparable to a basic building material like brick at 3 MJ/kg. Much the same can be said for embodied carbon, with outcomes between 0.0 and 0.5kg CO_2/kg, or closer together, again as shown in Figure 47. Yet another way of looking at these outcomes is by way of examining, material by material, how they comport to their embodied energy coefficients. As shown in Figure 48 in all but one case most materials align within the less than 5 MJ/kg category. This is not surprising as most basic building materials like concrete, brick, and timber are rated as less than 5MJ/kg. In addition, all cases have the presence, to a lesser or greater degree, of materials greater than 90 MJ/kg like aluminum, plastic, and rubber. The exceptional high case is Westwood in the U.S., representing the 'dispersed city' category. There it has a relatively low proportion of less than 5 MJ/kg coefficient materials and exhibits a broader and more complete distribution of materials across a fuller range of embodied energy coefficients. In fact, only 25 percent of the total embodied energy in the case is from less than 5 MJ/kg as shown in Figure 47. Understandably, it also has a higher proportion of brick buildings, not unusual for American single-family dwelling but with an embodied energy coefficient over three times higher than the more usual concrete in use elsewhere as the basic material. At the other end of the range, La Campana in Monterrey has a relatively high number of materials within the less than 5 MJ/kg category and with the majority of buildings made of concrete at less than or equal to 1.0 MJ/kg, but where a total of fully 50 percent of its total embodied energy comes from materials below 5 MJ/kg as well as much smaller proportions of higher coefficient materials like aluminum. Again, this is to be expected as the levels of finish and material complexity in informal or 'organic settlements' is low compared to elsewhere. Chencun has 38 percent of total embodied energy less than 5 MJ/kg, whereas the Paris ZAC outperforms the other cases environmentally which might be expected from its resource-conserving profile.

Moving still further into the data, other mitigating factors are the assumed obsolescence rates for the various cases and components within those cases, as shown in Figure 49. Again, as expected, both the Paris ZAC and Westwood are the most stable with regard to buildings, with roadways across all cases at much the same rates of 15 to 20 years and the same for vehicle fleets. In the latter, given the underlying economic circumstances of informal settlements, cars, for instance, are turned over less often than in more affluent circumstances, but building modifications occur more frequently in response to population dynamics and household affluence. In short, fundamental socio-economic differences among the definitions of the domains of settlement coexistence are directly related to and are reflected in these varying obsolescence rates, as might be expected, and the ultimate environmental performance involved.

In addition, waste management and the circularity of phases in the stock-flow modeling process also seem to follow suit. In more developed circumstances like the Paris ZAC and Westwood among disposal of collected waste, recycling is higher in Paris than in the American dispersed city and landfilling as a practice is slightly lower. In a final accounting, of the 16,299 metric tons of collected waste in the Paris ZAC, 9,036 tons is recycled mainly through material mining and 3,777 is land filled. At Westwood 68,096 metric tons is collected, 17,153 tons is recycled, and 29,688 metric tons is landfilled. Taking landfilling as perhaps the most problematic form of waste disposal, 23 percent of collected waste occurs in Paris, whereas 44 percent occurs in the dispersed circumstance of Westwood. La Campana has little recycling, though a local program is being launched. However, the higher obsolescence rates in buildings imply that the circularity of repair, refurbishment, and reuse are also likely to be higher than in the more stable and, one might say, sedentary circumstances. This is also borne out more generally, through the observation of higher rates of change and the flexibility that goes with that vis-a-vis daily life in self-organized informal or organic settlements. Waste management circumstances in China are underdeveloped though improving rapidly. Today some 50 percent of household waste is recycled and some 30 to 40 percent of total waste comes from construction and demolition activities, but with a very low recovery rate of only about 5 percent (Huang, Wang, and Kua, 2018).

FIG.50: Percentage Distribution of Sources of Water by Case.
FIG.51: Percentage Distribution of Water by Use and Case.
FIG.52: Percentage Distribution of Water Output by Case.

	Surface	Ground	Storm
Paris ZAC	23	23	54
Westwood	10	90	—
La Campana	62	38	—
Chencun	46	—	54

	Residential	Commercial	Industrial	Greenspace	Agriculture
Paris ZAC	39	36	1	24	—
Westwood	36	45	1	18	—
La Campana	83	11	—	6	—
Chencun	20	3	2	2	73

	Waste Water Treatment	Leakage	Evaporation	Infiltration
Paris ZAC	59	10	18	13
Westwood	64	5	14	17
La Campana	91	6	1	2
Chencun	28	9	28	35

With regard to the water stock-flow models, there were two major sources of supply as shown in Figure 50. They were surface water and ground water. In the Paris ZAC a sophisticated stormwater capture and management system was also installed as a part of the project's urban design, particularly around Martin Luther King Park. Elsewhere piped water systems were used predominantly, without much by way of stormwater harvesting, except at Chencun where much of the agricultural use drew on stormwater supplies as shown in Figure 50. Different characteristics were also reflected in the land use associated with water use. At La Campana it was mainly residential accounting for 83 percent, whereas in the Paris ZAC it was relatively evenly split at around 39 percent residential and a further 36 percent commercial and public use. Yet again in both Westwood and Chencun the preponderance of either private gardens or agriculture accounted for 18 and 73 percent of use respectively and are shown in Figure 51. Domestic water use via residential utilities was common in all cases and more or less to the same degree, depending upon residential population. The more efficient water model for La Campana relied almost solely on water-efficient appliances netting something like a 42 percent reduction in water demand. With regard to final output, they were of four basic kinds, namely wastewater treatment, leakage, and both evaporation and infiltration essentially from land areas associated with uses and as shown in Figure 52. There, the effect of these latter areas came into play in Westwood, mainly through private gardens, and in Chencun's agriculture, at around 30 percent of its total output. The impacts of these two cases on green space, an important aspect of settlement metabolism, are also likely to be more elaborate. Water recycling appears to be a possibility and is used, for instance, in the Paris ZAC, but could also yield up to around 50 percent return of gray or potable water in the other cases. Although Chencun is the only case that is in a largely water-stressed geographical context, the diminution in scale of water-related infrastructure accessioned by more comprehensive water recycling is likely to be beneficial metabolically on the other cases.

With regard to the energy stock-flow models, there were six singular sources. They were renewables, nuclear, natural gas, petroleum, coal, and others, including biomass and as shown

	Renewable	Nuclear	Natural Gas	Petroleum	Coal	Other
Paris ZAC	16	59	6	19	—	1
Westwood	21	—	41	36	—	2
La Campana	13	2	31	45	8	1
Chencun	8	15	13	10	52	2

	Residential	Industrial	Agriculture	Commercial	Transportation
Paris ZAC	25	4	—	33	38
Westwood	23	2	—	39	36
La Campana	43	—	—	12	45
Chencun	9	20	57	3	11

	Losses	Total Consumed
Paris ZAC	62	38
Westwood	62	38
La Campana	66	34
Chencun	64	36

FIG.53: Percentage Distribution of Energy Sources by Case.
FIG.54: Percentage Distribution of Energy by Land Use by Case.
FIG.55: Percentage Distribution of Energy Output by Case.

in Figures 53, 54, and 55. Put another way they are renewables, nuclear, and fossil fuels. Among the cases the Paris ZAC was mainly fueled by nuclear power at 59 percent of its total in keeping with France's move in that direction from the 1980s to the highest national proportion in the world at around 72 percent of its total. Westwood was mainly dependent on fossil fuels at 77 percent of its total and with 41 percent from natural gas and 36 percent from petroleum. La Campana was 84 percent dependent on fossil fuels, including petroleum at 45 percent and coal at 8 percent, whereas Chencun was 75 percent dependent on fossil fuels with a large 52 percent proportion in coal, like much of China these days. Among renewables Westwood had the highest proportion, mainly from solar, followed closely by the Paris ZAC at 16 percent and with La Campana and Chencun at 13 and 8 percent respectively. As far as use-related expenditure Westwood, Paris ZAC, and La Campana were private-automobile-dependent forms of settlement at 36, 38, and 45 percent of their totals respectively going to transportation and with Chencun, as a desakota area, with a substantial investment of 57 percent of its total in agriculture. The least auto-dependent settlement was Chencun with only 11 percent of their totals going into transportation, respectively. In keeping with basic laws of thermodynamics, outputs for all cases at over 60 percent in losses surpassed consumption. Onsite renewables from geothermal and solar installations occurred at the Paris ZAC and could occur elsewhere such as Westwood. There, overall gains from shifts to electric vehicles would need to also reflect source power generation less dependent on fossil fuels.

B. Specific Urban Projects.

In addition to these four domains of settlement co-existence which are seen here to be global in occurrence, there are particular urban projects and building types that warrant scrutiny in terms of metabolic performance. These comprise particular urban districts or urban design projects such as at the Champs-Élysées in Paris, as well as particular housing types typical to specific domains of settlement co-existence like, for example, American single-family housing in dispersed community settings or denser apartments in compact settlements such as those in

Paris, France. Then too, there are planned proposals for projects in developing situations, such as at the Victorias Eco-Industrial Park on Negros Island in the Philippines.

i. Champs-Élysées Commercial District

Described at least by some as the 'most beautiful avenue in the world' for over a century, the Avenue des Champs-Élysées has occupied a unique place in the international collective imagination (Herment, 1856). More blandly described, it is the avenue in the 8th Arrondissement of Paris, some 1.9 kilometers in length and around 70 meters in width, running between the Place de la Concorde in the east and the Place Charles-de-Gaulle replete with the Arc de Triomphe in the west. It is named in French for the Elysian Fields, and is subdivided more or less into two segments, with the lower part running from the Place de la Concorde in the east to the Rond-Point des Champs-Élysées at the center and as the area of the Jardin de Champs-Élysées containing the Grand Palais, the Petit Palais, and the Théâtre Marigny alongside restaurants, gardens, and other monuments. The upper part runs from the Arc de Triomphe, or L'Étoile, and is known for its cafes, restaurants, luxury stores, and other commercial establishments. In contrast to the lower portion, it is comprised of five- to six-story buildings flanked by broad sidewalks and by trees and with four lanes of traffic in each direction as shown in Figure 56. At least since the 1980s there has been strong growth in tourism, at the expense of local populations who now infrequently go there. Historically the Champs-Élysées and its gardens were originally laid out in 1667 by André Le Nôtre as an extension of the Tuileries Gardens to the east which had been built in 1564 and modified by Le Nôtre in 1664. Then in 1710 the avenue was extended beyond the Rond-Point des Champs-Élysées as far as the modern Place de l'Étoile and by the late 19th century the upper portion had become a fashionable avenue and part of the city by 1826, as shown in Figure 57, as well as a lower portion, shown in Figure 58 (Jarrasse and Lebar, 2007). The Arc de Triomphe was commissioned by Napoleon Bonaparte after Austerlitz in 1805 but remained unfinished until 1836. The avenue was also and continues to be the site of military parades. In 1860 merchants in the upper portion joined together to promote commerce. In 1980 the group

FIG.56: Aerial Oblique View of the Champs-Élysées.
FIG.57: Ground Level View of the Upper Champs-Élysées.

became the Comité des Champs-Élysées and simply the Comité Champs-Élysées in 2008. The group has perpetually dedicated itself to seeking public projects to enhance the avenue's unique atmosphere and to lobby for business hours and the organization of special events. In fact, a redevelopment project was led by the architect Bernard Huet in 1992–1994 to give more space to pedestrians, as shown in Figure 58, but did not manage to reverse the dynamics and problems of rising air pollution from some 3,000 vehicles per hour, causing a general downtrend in the quality of the district with the layout depicted in Figure 59. In 2019 the Comité Champs-Élysées decided to engage in a more fundamental reflection on the avenue and commissioned the architecture and urban design firm of PCA-STREAM to produce a strategic vision (PCA-STREAM, 2020; and PCA-STREAM, 2023) and as recently as 2021 Anne Hidalgo, the Mayor of Paris, began arguing for a plan to transform the Champs-Élysées into an 'extraordinary garden' with a substantial reduction in traffic. The current vision is largely defined by three design measures. The first is to reduce the road to a 2×2 lane affair of motorized traffic, as shown in Figure 60. The second is to increase the green permeable space within the avenue, and the third is to trigger some changes in adjacent uses especially in the direction of restructuring retail space and expanding office space to make the adjacent uses both more usable and attractive particularly for locals, as depicted in Figure 61. The site and programmatic components of this proposed district redevelopment in the upper portion is the subject of this metabolic modeling exercise.

These three aspects of redesign and redevelopment were considered and modeled (Doussard, Rowe, Detavernier, and Guo, 2024). The first entailed reduction of the motorized roadway from 4×4 to 2×2 lanes as detailed earlier. According to mobility consultants this would result in a 19 percent reduction in traffic without additional congestion. It also made way for additional layers of bike lanes and sidewalks. The second aspect was to increase the amount of permeable surface within the extended sidewalk areas. Currently 98 percent of the site surface is impermeable and paved in marble paving and cobblestones. Additional Stockholm tree planting pits were proposed along with some areas for shrubs and bushes. The third aspect concerned real-estate turnover within adjacent buildings, presumably

FIG.58: A View of the Lower Champs-Élysées.

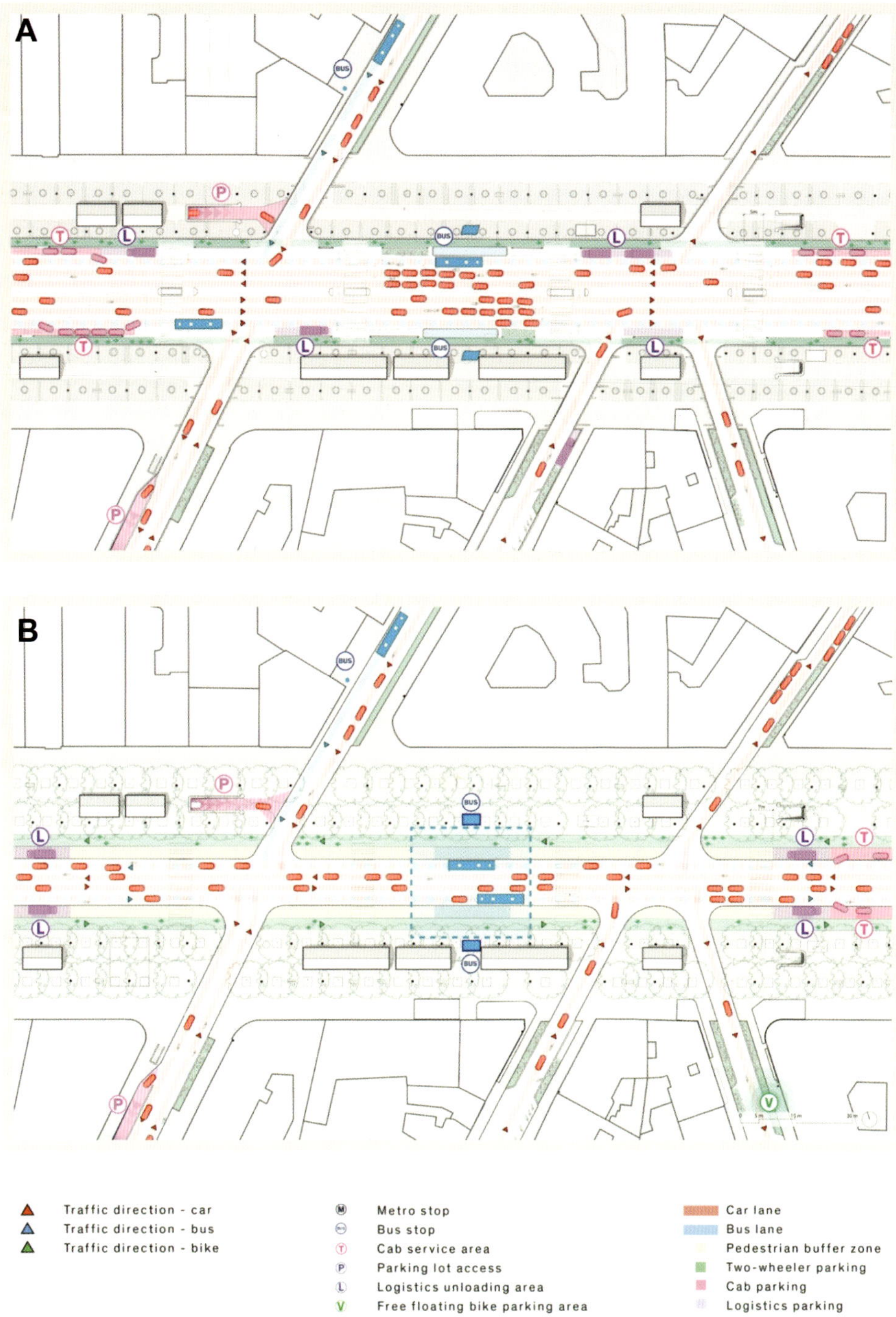

FIG.59: The Existing Upper Champs-Élysées 4×4 Mobility Plan.
FIG.60: The Proposed Upper Champs-Élysées 2×2 Mobility Plan.

encouraged by modification of the avenue. Local data indicated that real-estate turnover at the ground level had been significant during the last 10 years, and it was expected to continue. Reliable data gathering was an issue and was approached in four ways. First, data was provided by PCA-STREAM in a precise manner regarding existing avenue conditions and likely future conditions, especially for the change from 4×4 to 2×2 motorized lane configurations as described earlier in Figures 59 and 60. Second, data was provided by local sources, such as APUR or Atelier Parisien d'Urbanisme, the City of Paris, and the Ministry of Ecological Transition. Third, interviews were conducted with commercial space tenants, with sampling among them that was representative of their presence along the avenue. Finally, in the absence of availability from these three local sources, regional and national authority reports were consulted. Overall, however, almost all of the data came from site-specific and local sources.

The primary method of analysis, as described in Chapter 3 of this volume characterizing stock-flow modeling, involved mass conservation principles in which the mass of inputs in a process was equal to the mass of outputs plus a storage term. In addition, as elsewhere three separate models were developed dealing with water, energy, and material flows. Consideration in these regards was associated with commercial activities adjacent to the avenue, public spaces along the avenue, and transportation associated with the flow of vehicles along and through the avenue, such as bikes, cars, vans, trucks, and buses. Results of each modeling exercise again as elsewhere in this volume were represented in the form of Sankey diagrams and associated accounts of embodied energy and carbon using appropriate coefficients.

Results for the water model demonstrated that most of the water consumed was associated with adjacent building activities, in which restaurants and offices were the most consumptive as shown in Figure 62. By contrast, stocks and flows associated with public spaces were negligible but with leakage at 20 percent as significant, no doubt owing to the age and state of piped infrastructure. Primarily two aspects of proposed redevelopment appeared to have the most impact on water processes. They were, first, much the same proportions of evaporation, infiltration, and runoff associated with increases in permeable surfaces. Second, were likely increases in water consumption from the

turnover of adjacent uses, especially from restaurants and food courts. The energy model, depicted in Figure 63, showed that most energy consumption was due to adjacent commercial activity and largely in the form of electrical use. Transportation only represented some 3.5 percent of all end-use consumption and mainly in the form of petroleum use. Consumption associated with public space was negligible. Materially, most of the stocks and flows were associated with transiting along and through the avenue and with shopping. Indeed, as shown in Figure 64, 95 percent of all materials were associated with this transiting

FIG. 61: Part of the PCA-STREAM Vision of the Champs-Élysées.

Champs-Élysées WATER MODEL

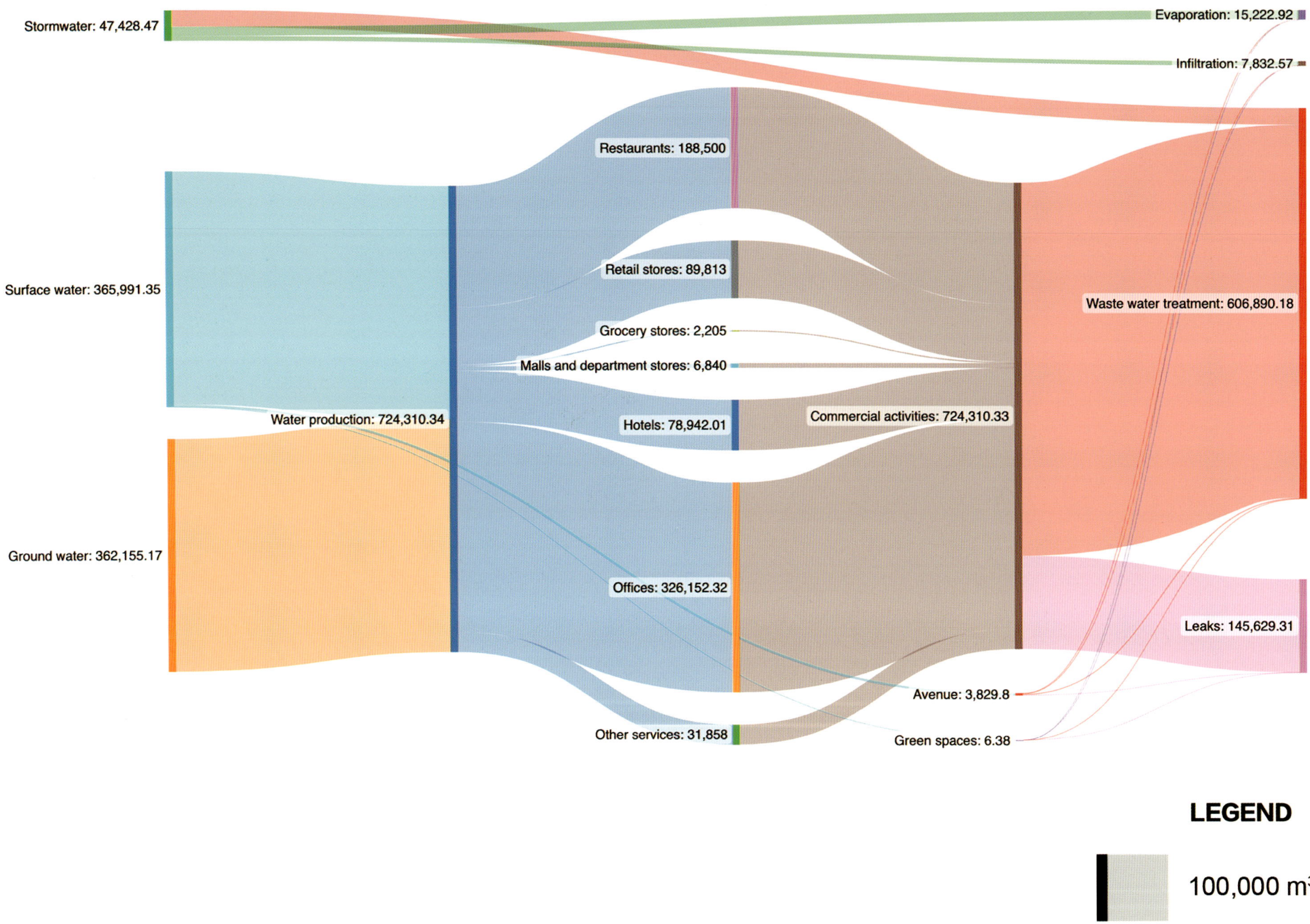

FIG.62: Champs-Élysées Water Sankey Diagram

Champs-Élysées ENERGY MODEL

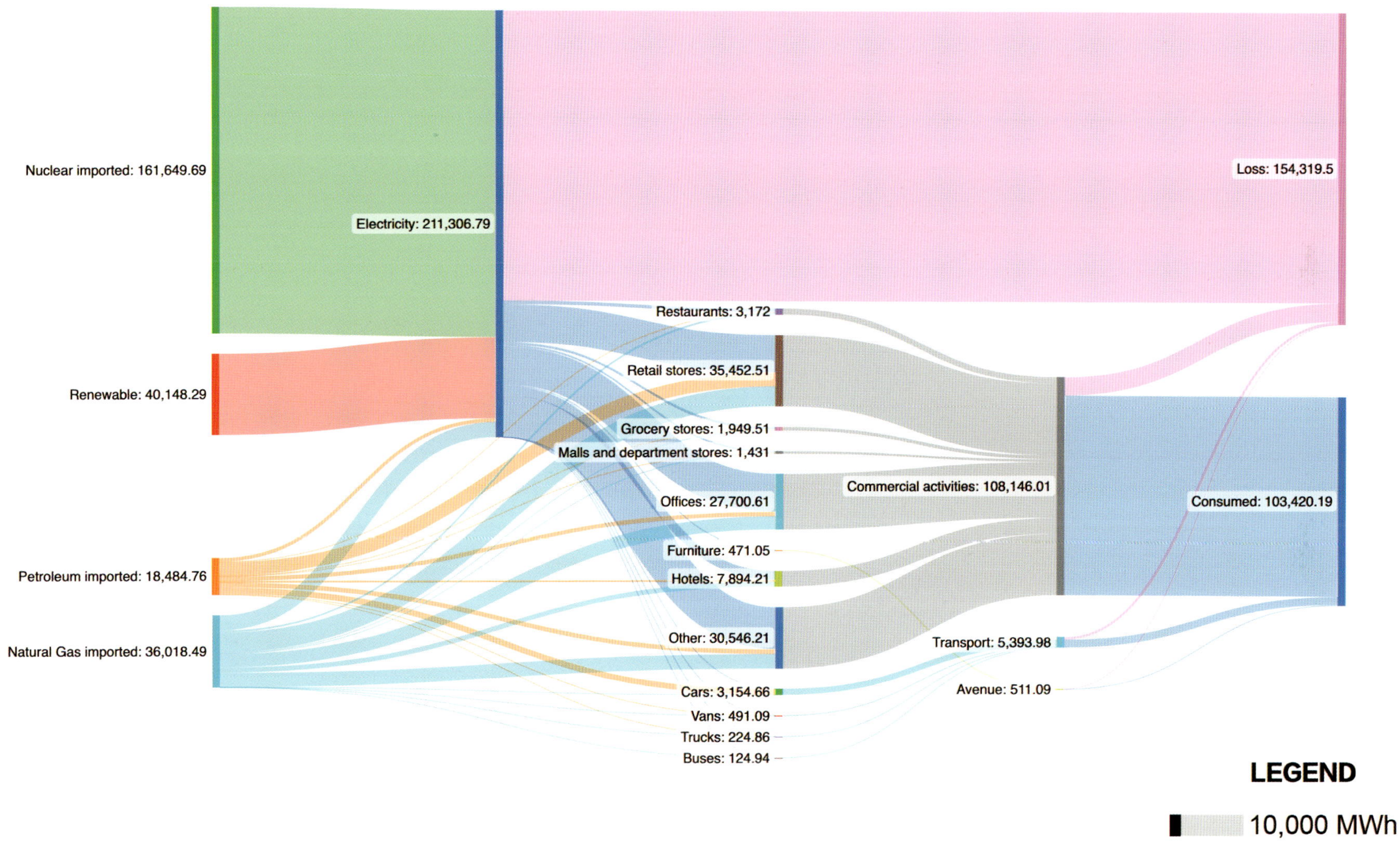

FIG.63: Champs-Élysées Energy Sankey Diagram

Champs-Élysées MATERIAL MODEL

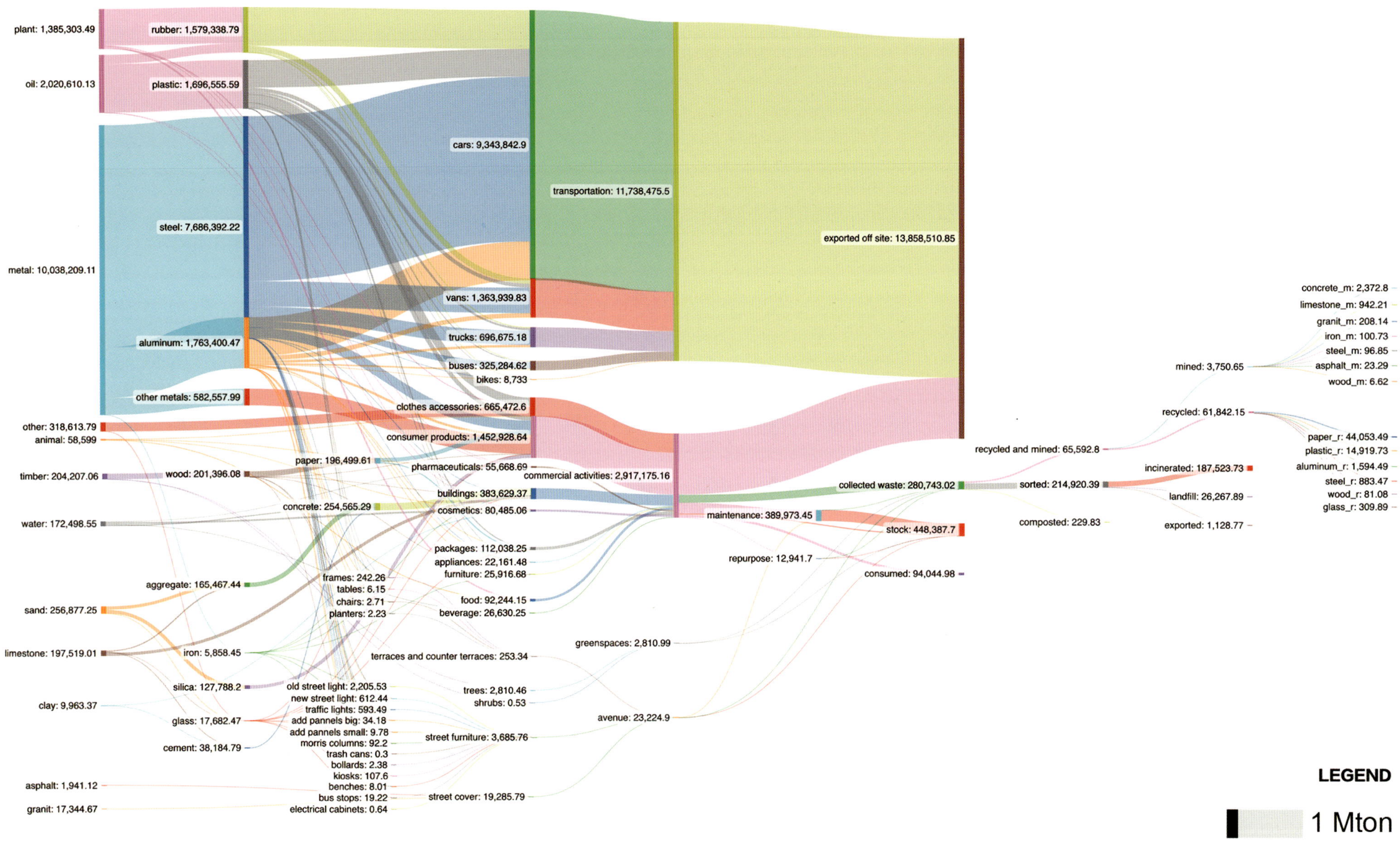

FIG. 64: Champs-Élysées Material Sankey Diagram

and with 80 percent of that in transportation, of which most was materially in metal such as the vehicles themselves, as well as in high-tech goods, consumer products, and street furniture. Actual on-site consumption due to commercial activities represented some 20 percent of all stocks and flows but the share associated with building and infrastructure remained much the same between current and redeveloped circumstances. In addition, waste production as modeled was low at less than 2 percent of total weight and with about 22 percent either mined or recycled. Overall, redesign or redevelopment does not appear likely to yield significantly reduced environmental impacts. This also includes the accounting of embodied energy in MJ/year associated via appropriate coefficients (Doussard, Rowe, Detavernier, and Guo, 2024).

ii. Housing Types.

Among the single-family housing types that dominate residential development within the peripheral kind of settlement co-existence in Westwood, Massachusetts, 21 Strasser Avenue was selected for detailed analysis. It is reasonably typical, with a capacious living area of some 274 square meters across an upper and lower level and with an oversized two-car garage of a further 60 square meters. Built in 1959 and shown in Figure 65, it is generally identifiable as a raised ranch-style house, typically of wood-frame construction, a concrete foundation, and shingle-tiled roof (Rowe, 1991). Specific building statistics are tabulated in Figure 66. The site area is around 0.2 hectares and includes a heated swimming pool and wooden deck area in the backyard, as shown in Figure 67. It is also ideally situated for family-style living next to Downey Elementary School and the property has been updated to accommodate handicapped users. Cooling is supplied by a central air conditioning system and heating via baseboard units using natural gas. The interior of seven rooms has a large eat-in kitchen, as shown in Figure 68. It also includes a sizable dining area, family room, primary bedroom, and bathroom, with two additional bedrooms with bathrooms.

The material weight of the house is around 274 metric tons per year with a lifespan of 61 years, including relatively constant upkeep and maintenance, as depicted in Figure 69 through the

Building Information

Field	Description
Year Built	1959
Living Area	1,788
Replacement Cost	$493,394
Building Percent Good	75
Replacement Cost Less Deprecation	$370,000

Building Attributes

Field	Description
Style	Raised Ranch
Model	Residential
Grade	Average - Good
Stories	1
Occupancy	1
Exterior Wall 1	Clapboard
Exterior Wall 2	N/A
Roof Structure	Gable/Hip
Roof Cover	Asph/ F Gls/Cmp
Interior Wall 1	Plastered
Interior Wall 2	N/A
Interior Floor 1	Hardwood
Interior Floor 2	Carpet
Heat Fuel	Gas
Heat Type	Hot Water
AC Type	Central
Total Bedrooms	3 Bedrooms
Total Bathrooms	3 Bathrooms
Total Half Baths	N/A
Total Rooms	7
Bath Style	Average
Kitchen Style	Semi Modern
Number of Kitchens	1
Garage	2
Fireplaces	2
Chimney Type	Masonry
Basement	1

FIG.65: Image of 21 Strasser Avenue, Westwood.
FIG.66: Table of Building Statistics for 21 Strasser Avenue, Westwood.

material model Sankey diagram. Inputs from the geosphere account for about 54 percent of the total weight, followed by the biosphere and particularly the use of timber at 42 percent, and 3 percent from the hydrosphere. A conspicuous and dominant aspect of this material weight is the roadway component of building at 43 percent and as shown in Figure 69, with the residential building itself accounting for about the same amount. The annual fixed asset stock is 94 percent, with a further 4 percent in collected waste and around 2 percent in consumption. Of the collected waste 23 percent is recycled, primarily involving material mining; some 9 percent is composted; and 68 percent is sorted, with 57 percent of that being landfilled. Within the asset stock a large proportion is due to maintenance activity, as noted earlier.

With regard to the Parisian Haussmann apartments, an example was chosen at 1 Boulevard du Sebastopol at the border of the 1st and 2nd arrondissements, or districts of Paris, and those of the 3rd and 4th arrondissements, starting at the Place du Chatelet, as shown in Figure 70 (Benoit, 2017). It was opened up by the Baron Haussmann during his transformation of Paris in the 1850s and named to celebrate Napoleon III's victory at the port of Sebastopol in the Crimea of 1855. It is lined along much of its 1.3-kilometer length by typical five-or-so-story Haussmann blocks primarily of apartments on top of commercial uses. The particular unit under consideration here was designed by Lecomte. As shown in the standard apartment floor plan in Figure 71, entrance to a relatively compact unit of about 113 square meters occurs via a grand staircase onto an antechamber, with a kitchen towards the back of the unit also served by a separate service staircase. Bedrooms and, often, an office are arranged along the glazed front of the unit, with an open courtyard towards the back, providing both light and air into the L-shaped plan of the overall unit, again as shown by Figure 71 and the aerial view of the larger block depicted in Figure 72. The predominant materiality of the block is limestone, often quite thick with blocks on the order of 50 centimeters in depth and architecturally well composed into striking neo-classical facades as depicted in Figure 73.

The material model of each unit is some 292 metric tons per year and is represented by the Sankey diagram displayed in Figure 74, somewhat comparable in weight to the more spacious

FIG.67: Backyard Swimming Pool and Deck of 21 Strasser Avenue, Westwood.
FIG.68: Large Eat-in Kitchen of 21 Strasser Avenue, Westwood.

WESTWOOD UNIT MATERIAL MODEL
Usage: **274 ton/year**
Area: **274.3 sq m**
Occupancy: **3**
Life span of unit: **61 years**

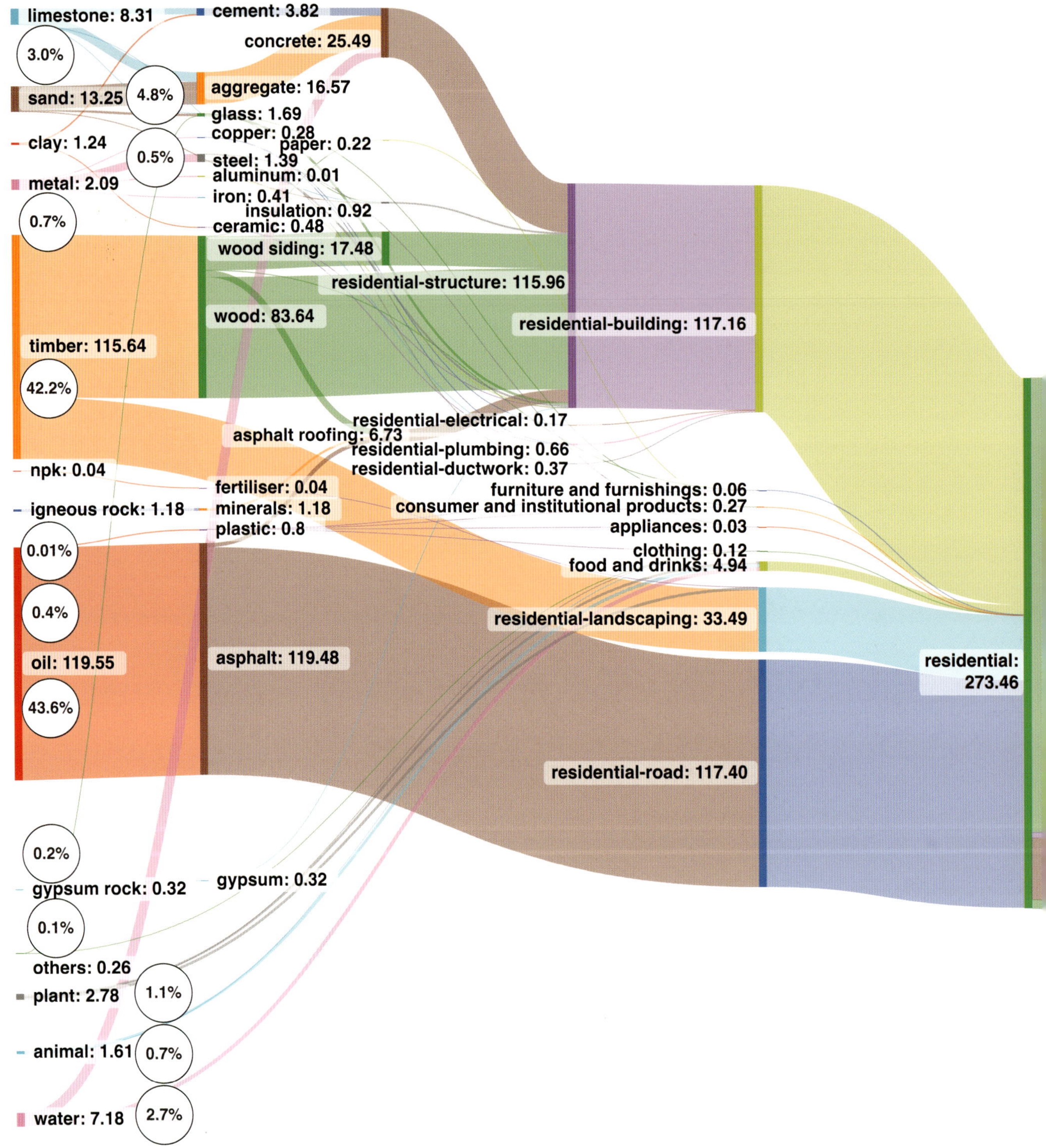

FIG.69: Material Flow Model of 21 Strasser Avenue, Westwood.

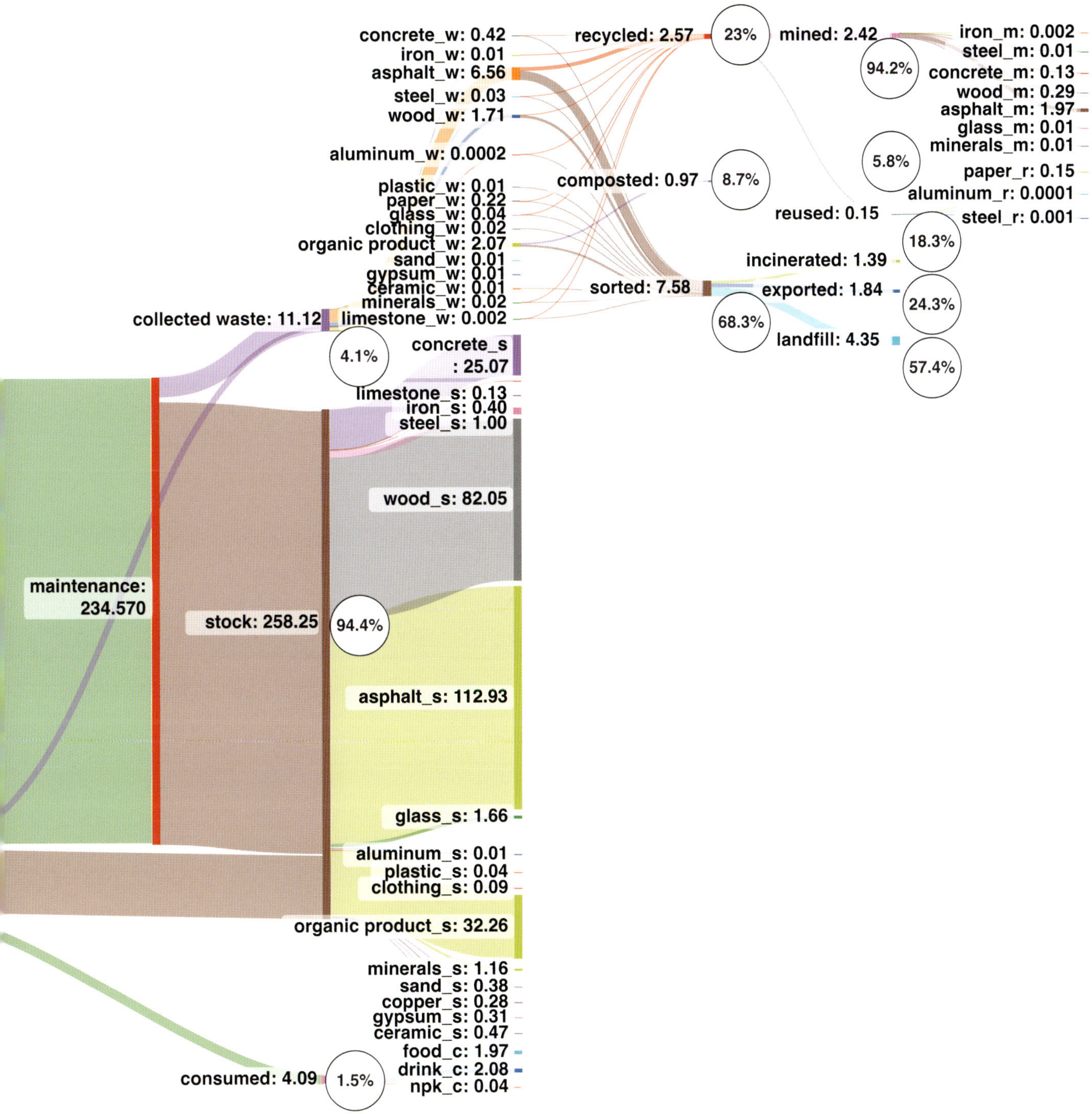
concrete_w: 0.42
iron_w: 0.01
asphalt_w: 6.56
steel_w: 0.03
wood_w: 1.71
aluminum_w: 0.0002
plastic_w: 0.01
paper_w: 0.22
glass_w: 0.04
clothing_w: 0.02
organic product_w: 2.07
sand_w: 0.01
gypsum_w: 0.01
ceramic_w: 0.01
minerals_w: 0.02
limestone_w: 0.002
collected waste: 11.12
4.1%
recycled: 2.57
23%
composted: 0.97
8.7%
sorted: 7.58
68.3%
mined: 2.42
94.2%
5.8%
reused: 0.15
incinerated: 1.39
18.3%
exported: 1.84
24.3%
landfill: 4.35
57.4%
iron_m: 0.002
steel_m: 0.01
concrete_m: 0.13
wood_m: 0.29
asphalt_m: 1.97
glass_m: 0.01
minerals_m: 0.01
paper_r: 0.15
aluminum_r: 0.0001
steel_r: 0.001
maintenance: 234.570
stock: 258.25
94.4%
concrete_s : 25.07
limestone_s: 0.13
iron_s: 0.40
steel_s: 1.00
wood_s: 82.05
asphalt_s: 112.93
glass_s: 1.66
aluminum_s: 0.01
plastic_s: 0.04
clothing_s: 0.09
organic product_s: 32.26
minerals_s: 1.16
sand_s: 0.38
copper_s: 0.28
gypsum_s: 0.31
ceramic_s: 0.47
consumed: 4.09
1.5%
food_c: 1.97
drink_c: 2.08
npk_c: 0.04

FIG. 70: Location of 1 Boulevard de Sebastopol, Paris.

1, BOULEVARD DE SEBASTOPOL

ARCHITECT- **LECOMTE**

FIG.71: Standard Floor Plan at 1 Boulevard de Sebastopol, Paris.
FIG.72: Aerial View of the Block at 1 Boulevard de Sebastopol, Paris.
FIG.73: Limestone – Facades at 1 Boulevard de Sebastopol, Paris.

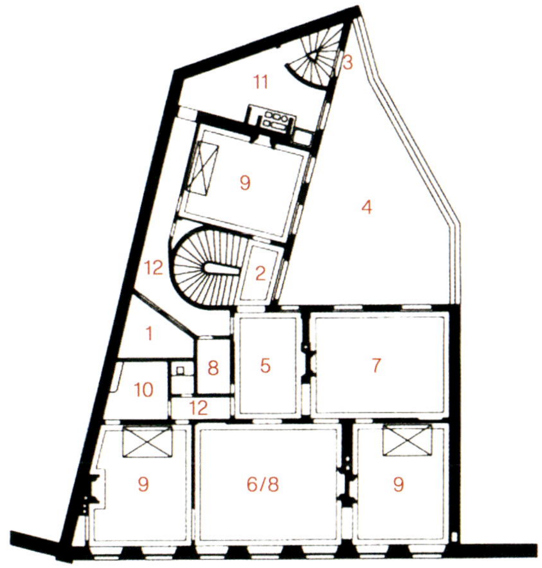

STANDARD FLOOR

1. Courtyard
2. Grand Staircase
3. Service Staircase
4. Courtyard
5. Antechamber
6. Living Room
7. Dining Room
8. Office
9. Bedroom
10. Toilets
11. Kitchen
12. Corridor

0 10 M

PARIS UNIT MATERIAL MODEL
Usage: **292 ton/year**
Area: **112.7 sq m**
Occupancy: **4**
Life span of unit: **200 years**

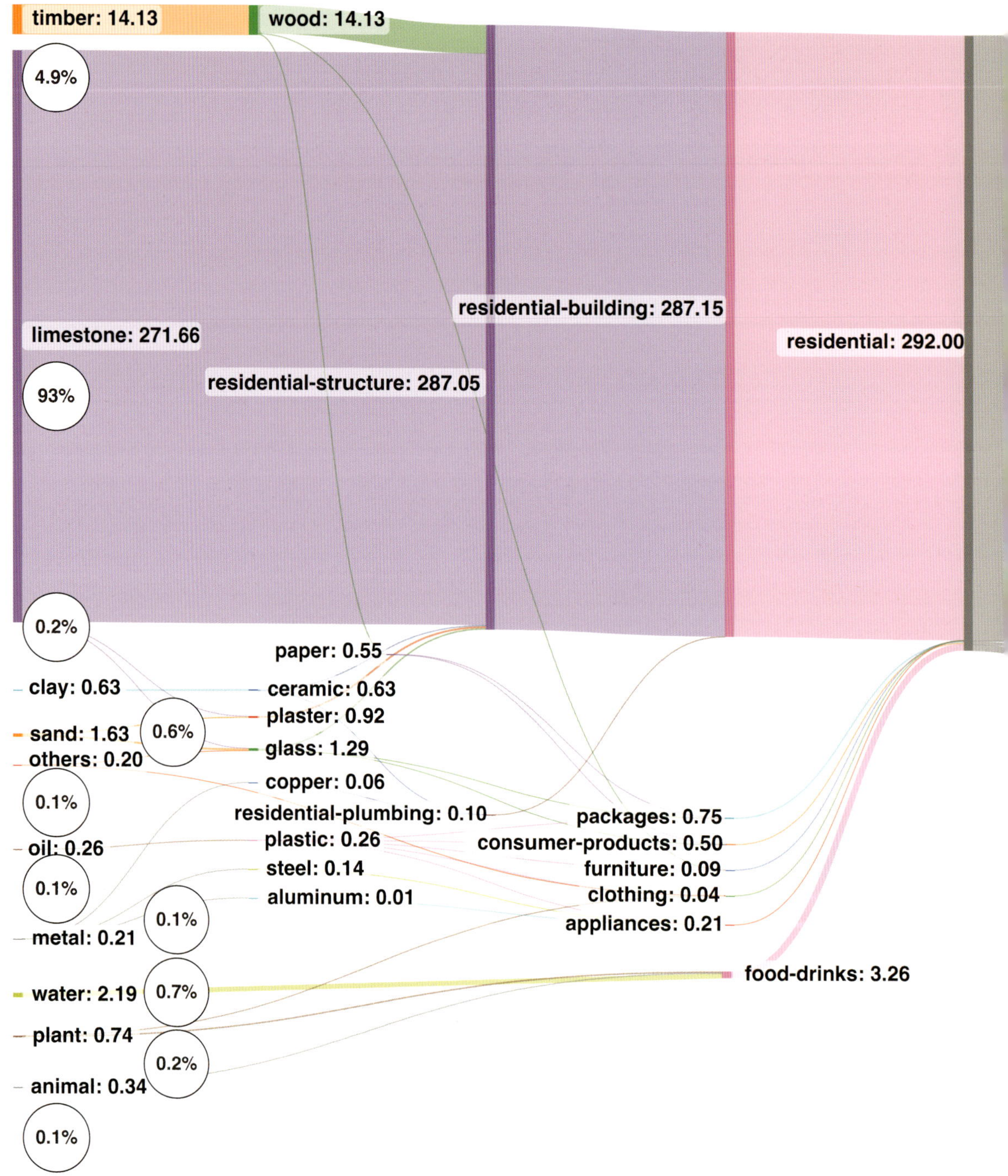

FIG. 74: Sankey Diagram of the Paris Unit Model.

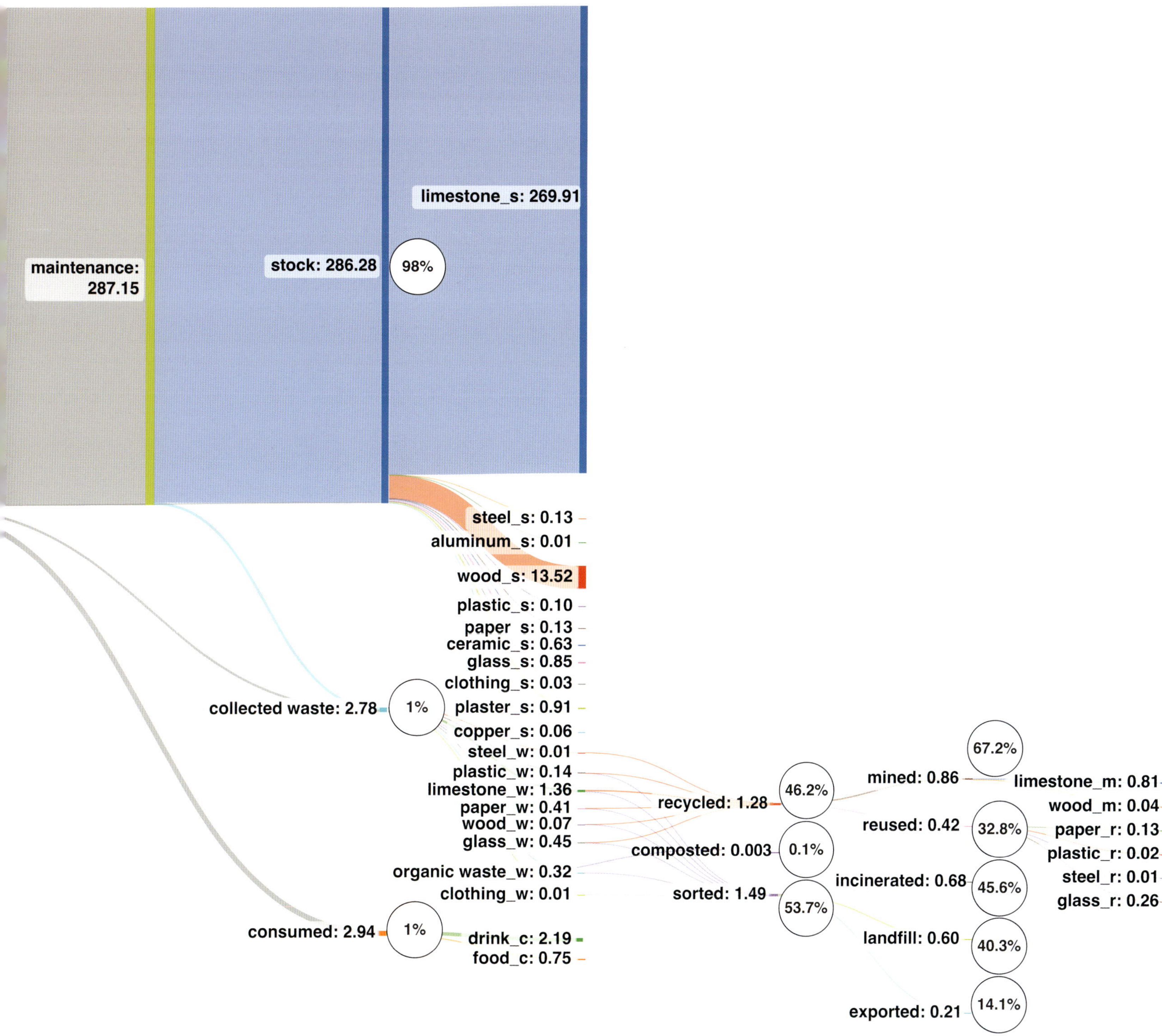

maintenance: 287.15
stock: 286.28
98%
limestone_s: 269.91
steel_s: 0.13
aluminum_s: 0.01
wood_s: 13.52
plastic_s: 0.10
paper_s: 0.13
ceramic_s: 0.63
glass_s: 0.85
clothing_s: 0.03
collected waste: 2.78
1%
plaster_s: 0.91
copper_s: 0.06
steel_w: 0.01
plastic_w: 0.14
limestone_w: 1.36
paper_w: 0.41
wood_w: 0.07
glass_w: 0.45
organic waste_w: 0.32
clothing_w: 0.01
consumed: 2.94
1%
drink_c: 2.19
food_c: 0.75
recycled: 1.28
46.2%
composted: 0.003
0.1%
sorted: 1.49
53.7%
mined: 0.86
67.2%
reused: 0.42
32.8%
incinerated: 0.68
45.6%
landfill: 0.60
40.3%
exported: 0.21
14.1%
limestone_m: 0.81
wood_m: 0.04
paper_r: 0.13
plastic_r: 0.02
steel_r: 0.01
glass_r: 0.26

single-family home in Westwood, Massachusetts, described earlier. In keeping with the predominance of limestone and related materials, 94 percent of the material weight is drawn from the geosphere, with only 5 percent from the biosphere, and a nominal proportion from the hydrosphere. The stock asset is 96 percent and durable with a long obsolescence rate of some 200 years. Collected waste accounts for only around 1 percent of the total and consumption is at much the same level. Of the collected waste some 46 percent is recycled and usually mined or reused. About 54 percent is sorted, and 40 percent of this is landfilled, again as depicted in Figure 74 and somewhat more environmentally friendly than Westwood.

Comparatively of similar total material weight with 292 metric tons in Paris and 272 metric tons at Westwood, the difference in unit spaciousness is striking with 113 square meters in the former and 274 square meters in the latter. This also translates into an average of about 28 square meters per person and 73 metric tons per person in Paris compared to the much larger 91 square meters per person and 91 metric tons per person in the U.S. example. Again, as in the comparable neighborhood analyses, the difference in embodied energy and carbon favors the Parisian example. All in all, the sheer scale of material use in the Westwood residential unit development is larger and one might say more wasteful than the Parisian case. Operationally, the private car orientation of the American peripheral example is also strikingly poor in environmental performance.

iii. Victoria's New Government Center

The Victorias New Government Center (VNGC) is set within a new district. Not unlike traditional new towns that featured a government building as a locus of activity, the VNGC is meant to set the tone of development. The public square, or plaza, surrounded by civic buildings is here replaced by a series of landscape zones with specific roles in the city's water management plan.

The humanist civic ideal captured so eloquently by Urbano's perspective of an open common framed by architecture is replaced here by constructed nature: a tree nursery, a riparian corridor, a flood-retention forecourt, and cistern performance discs are the new forms of public space. They don't contain space or frame views centered on a human occupation. They are designed in partnership with the goals of channeling, storing, and gradually moving water across a field. They are flow spaces. They only make sense because they connect to the movement of water across a territory.

From the southern mountain range to the bay in the north, the district is designed as both an aqueduct and a lake. It is meant to hold and to move water simultaneously. Each road is conceived as a fluvial artery in a network. The human circulation infrastructure is therefore shared with the ecological flows of natural systems like runoff, seasonal typhoons, and migrating birds.

The VNGC is set at the nominal center of these landscape spaces. It is square in plan and almost perfectly symmetrical from both the horizontal and vertical planes. Its proportions and shape make it look like an oversized house or an agricultural shed overgrown by plant life. The aesthetic mix-ups are intentional. The island of Negros is famous for both its large hacienda homes sitting amidst sugarcane plantations and agro-industrial complexes with colossal chimneys spewing gases. The VNGC is meant to represent the possibility of domesticating the global pressures behind collective farming and food safety.

It might be a futile attempt. The forces at play within this space cannot be surely addressed at a building scale. The supply chains, financial networks, legal systems, and so forth overwhelm comprehension. How could we possibly make sense of this with architecture? And yet, buildings are both monikers and machines. They represent ideas while also actively engaging with places into which they are introduced. The VNGC is designed to operate at both these frequencies.

Floating above the VNGC is a quadrant of solar panels. Hanging from every other floor plate are planter rings encircling the building, providing home to a native shrub garden in the air, and protruding from the top, like those industrial flues, are heat-stacking ventilation stacks and water-catchment umbrellas. Together, these extra-large sub-assemblies become the building's aesthetic register. They are all pieces of a machine that connect to something else, yet the full story of their interrelation remains invisible to the viewer.

FIG. 75: Image of the Victorias New Government Center.

Part of this register involved poles aligned like sugarcane or bamboo stalks, evoking the geographical circumstances of the building, as shown in Figures 75 and 76. At issue, however, was the materiality of the poles. At first aluminum was considered as being light weight and durable. But with an embodied energy of around 200 MJ/Kg it was quickly replaced by steel at around 10 MJ/Kg. Then the next obvious choice was bamboo or sugarcane stalks themselves, with embodied energy around 1 MJ/Kg. Nevertheless, modeling, as depicted in Figure 77, showed that the shorter obsolescence rate of bamboo of around 20 years, compared to 75 years, resulted in a longer-term increase in embodied energy over steel. Counter intuitive though this sequence of choices may seem, it demonstrates the need for more complete material metabolic analysis.

iv. La Salle Chapel

In June 1991, at 1:42 in the afternoon, a volcanic event on a tropical archipelago changed the weather patterns across the entire globe. the eruption of Mount Pinatubo was the second-largest volcanic cataclysm of the 20th century. Bursts of gas-charged magma exploded into umbrella ash clouds that filled the sky for thousands of miles. Satellites tracked the ash cloud as it traveled several times around the globe. A typhoon arrived that same day, bringing high winds and mixing rain with ash into a material known as lahar that descended upon towns, like fluid cement entombing entire communities as far as a hundred miles from the crater. Over two million people were adversely affected, and the Philippines suffered $700 million of damage loss of economic output that year.

Three decades after the cataclysm of Mount Pinatubo, farmers have replanted their plots, communities have rebuilt their towns, and a botanical garden is being built along the banks of one of the rivers flowing from the crater. This will be the first research garden dedicated to regenerative plant growth. It is being led by one of the largest local university groups in the world, De La Salle, and will examine and exhibit post-cataclysmic landscapes in transition.

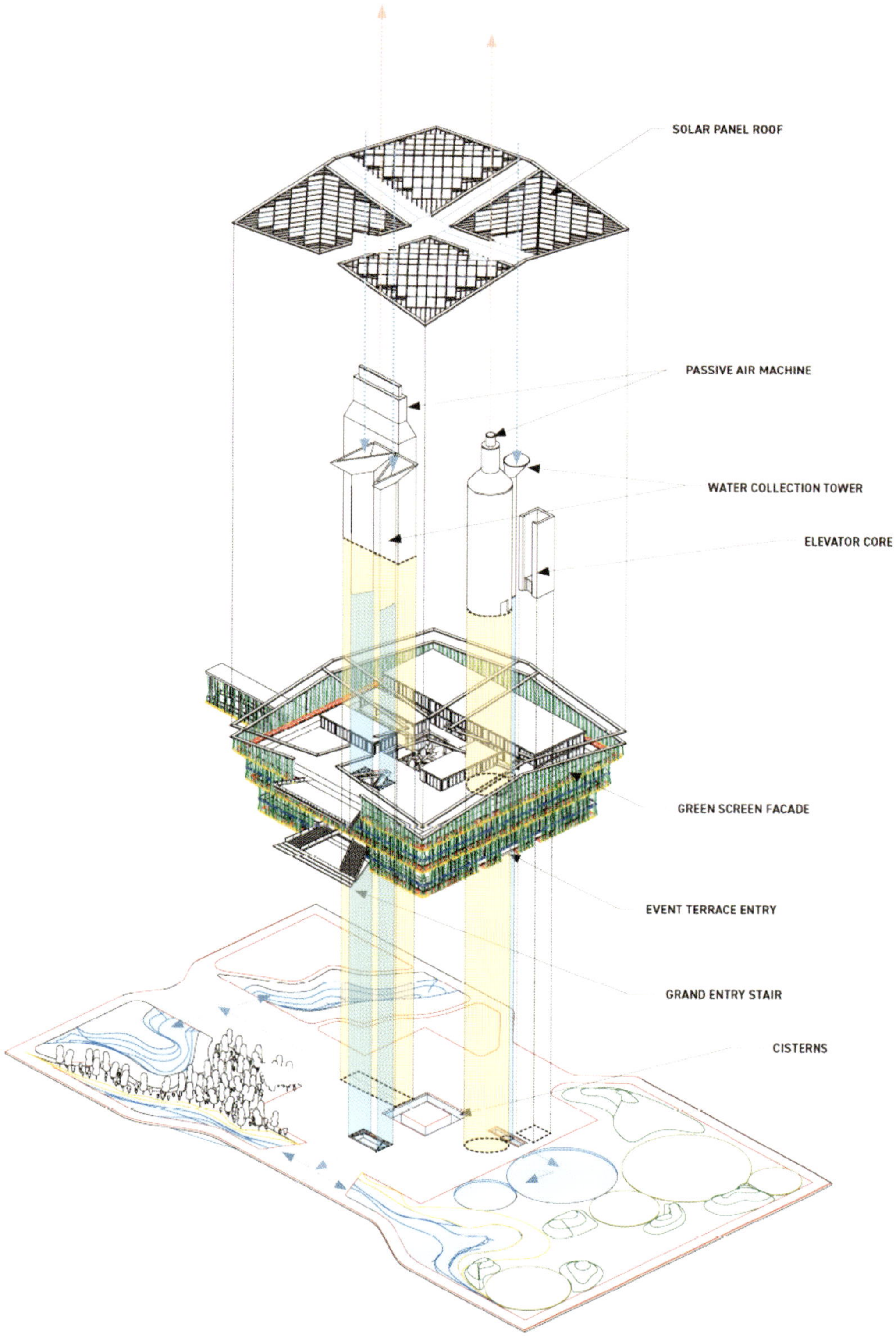

FIG.76: The Aesthetic Register of Building Assemblies.

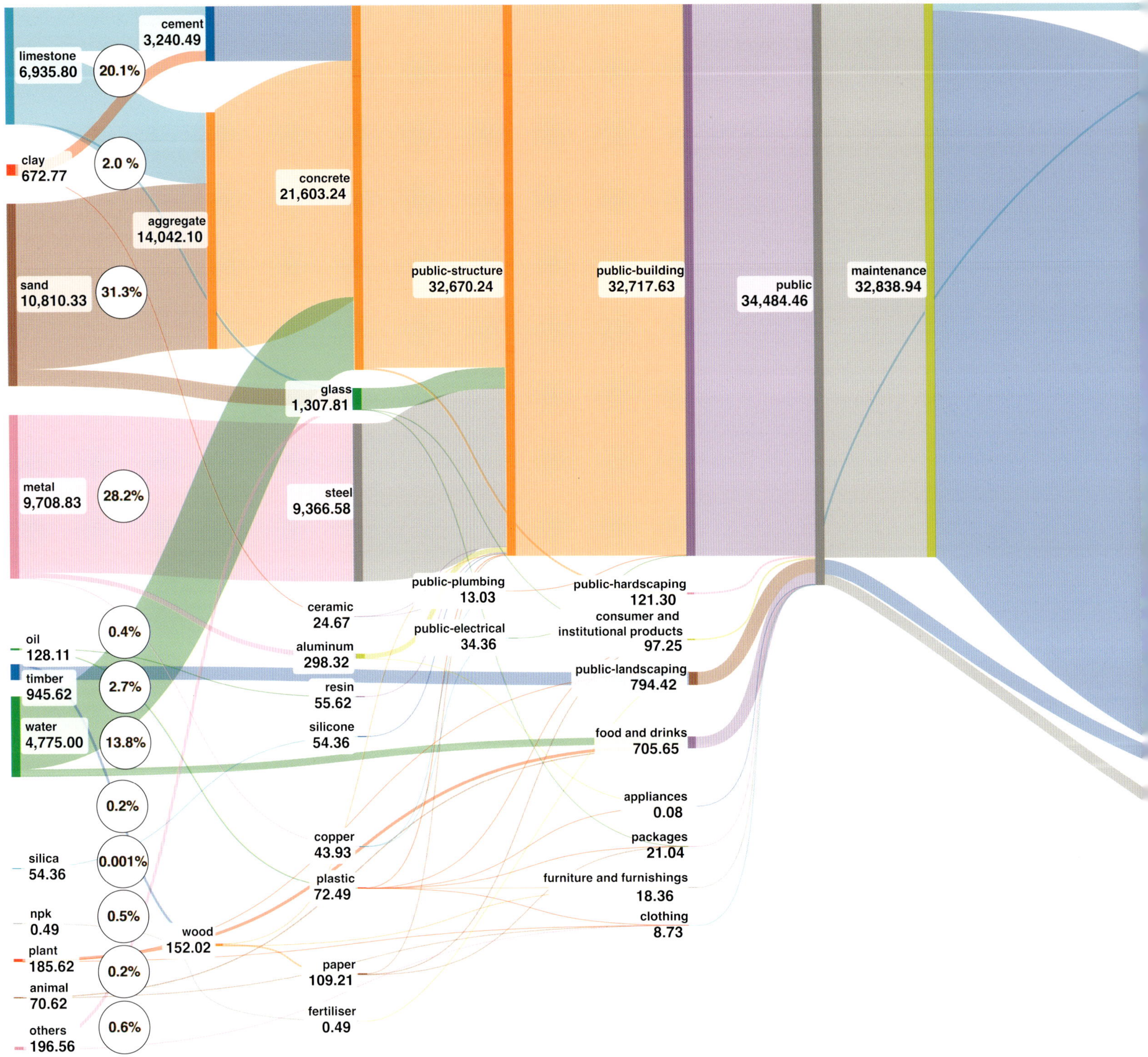
cement
3,240.49
limestone
6,935.80
20.1%
clay
672.77
2.0 %
aggregate
14,042.10
concrete
21,603.24
sand
10,810.33
31.3%
public-structure
32,670.24
public-building
32,717.63
public
34,484.46
maintenance
32,838.94
glass
1,307.81
metal
9,708.83
28.2%
steel
9,366.58
public-plumbing
13.03
public-hardscaping
121.30
ceramic
24.67
consumer and
institutional products
97.25
oil
128.11
0.4%
aluminum
298.32
public-electrical
34.36
public-landscaping
794.42
timber
945.62
2.7%
resin
55.62
water
4,775.00
13.8%
silicone
54.36
food and drinks
705.65
0.2%
appliances
0.08
silica
54.36
0.001%
copper
43.93
packages
21.04
plastic
72.49
furniture and furnishings
18.36
npk
0.49
0.5%
wood
152.02
clothing
8.73
plant
185.62
0.2%
paper
109.21
animal
70.62
others
196.56
0.6%
fertiliser
0.49

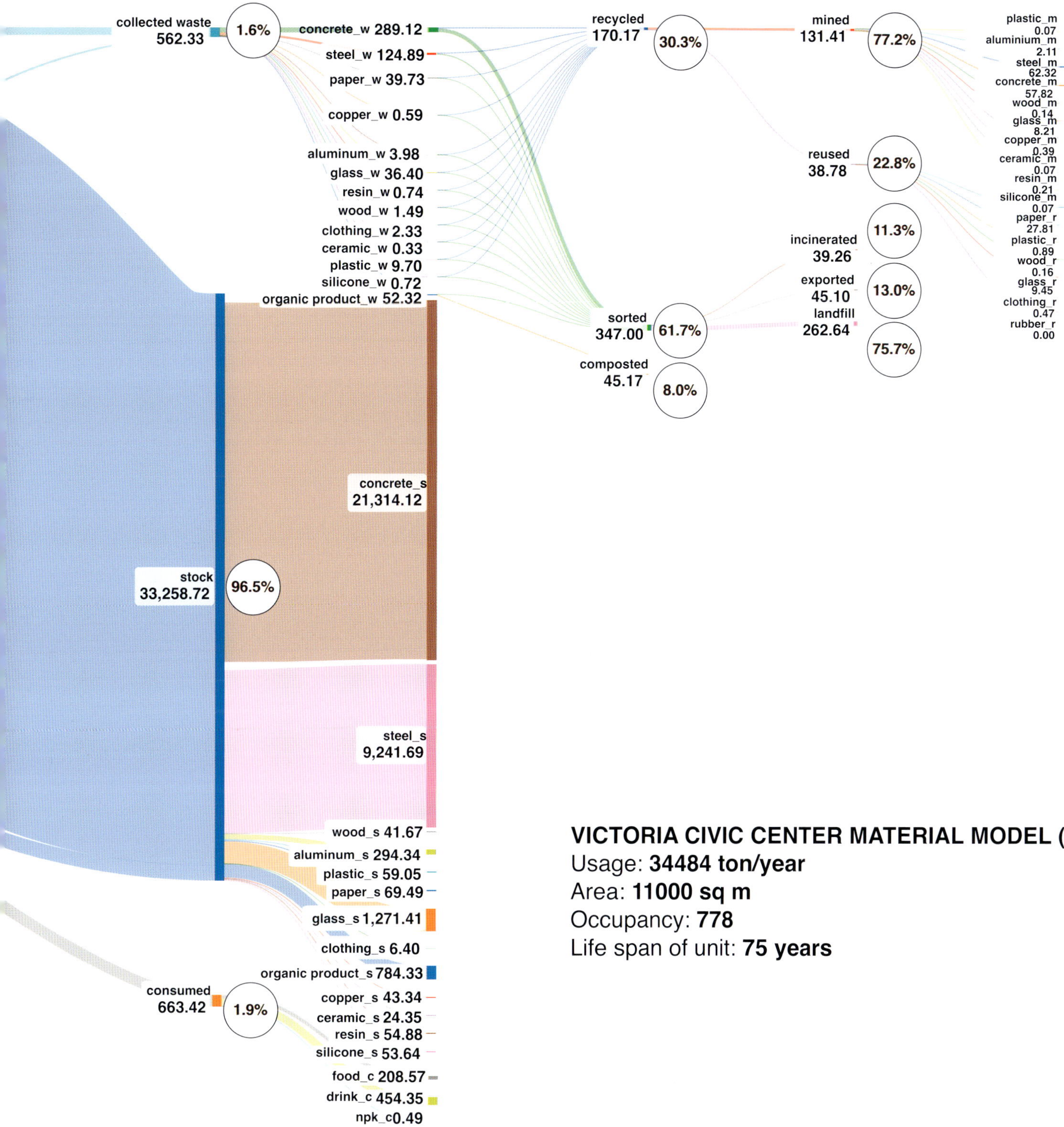

VICTORIA CIVIC CENTER MATERIAL MODEL (with aluminium facade and steel railing)

Usage: **34484 ton/year**

Area: **11000 sq m**

Occupancy: **778**

Life span of unit: **75 years**

FIG.77: Sankey Material Flow Model of the Center.

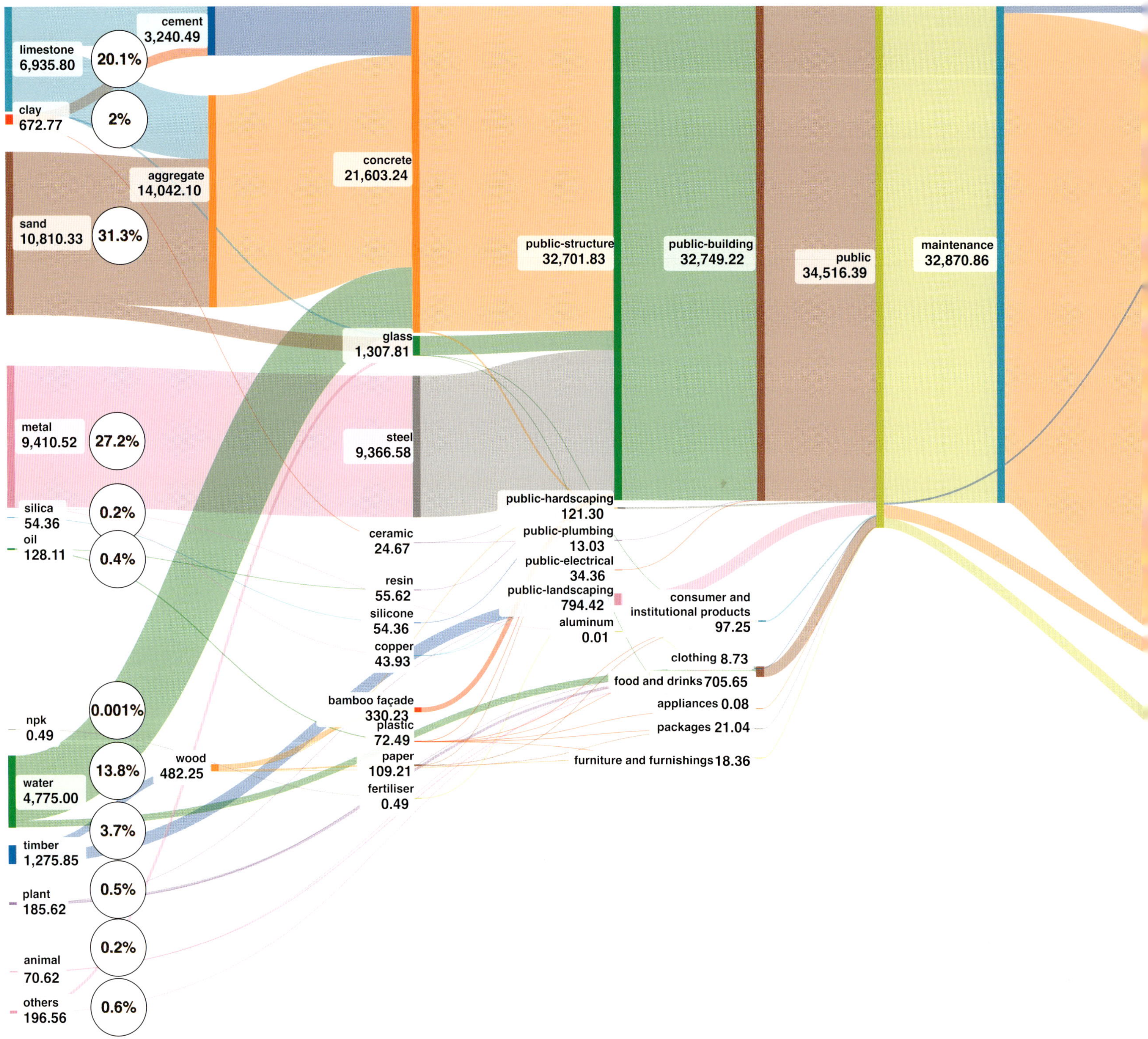

limestone
6,935.80
20.1%
cement
3,240.49
clay
672.77
2%
aggregate
14,042.10
sand
10,810.33
31.3%
concrete
21,603.24
glass
1,307.81
metal
9,410.52
27.2%
steel
9,366.58
silica
54.36
0.2%
oil
128.11
0.4%
ceramic
24.67
resin
55.62
silicone
54.36
copper
43.93
bamboo façade
330.23
plastic
72.49
paper
109.21
fertiliser
0.49
npk
0.49
0.001%
water
4,775.00
13.8%
wood
482.25
timber
1,275.85
3.7%
plant
185.62
0.5%
animal
70.62
0.2%
others
196.56
0.6%
public-structure
32,701.83
public-building
32,749.22
public
34,516.39
maintenance
32,870.86
public-hardscaping
121.30
public-plumbing
13.03
public-electrical
34.36
public-landscaping
794.42
aluminum
0.01
consumer and institutional products
97.25
clothing 8.73
food and drinks 705.65
appliances 0.08
packages 21.04
furniture and furnishings 18.36

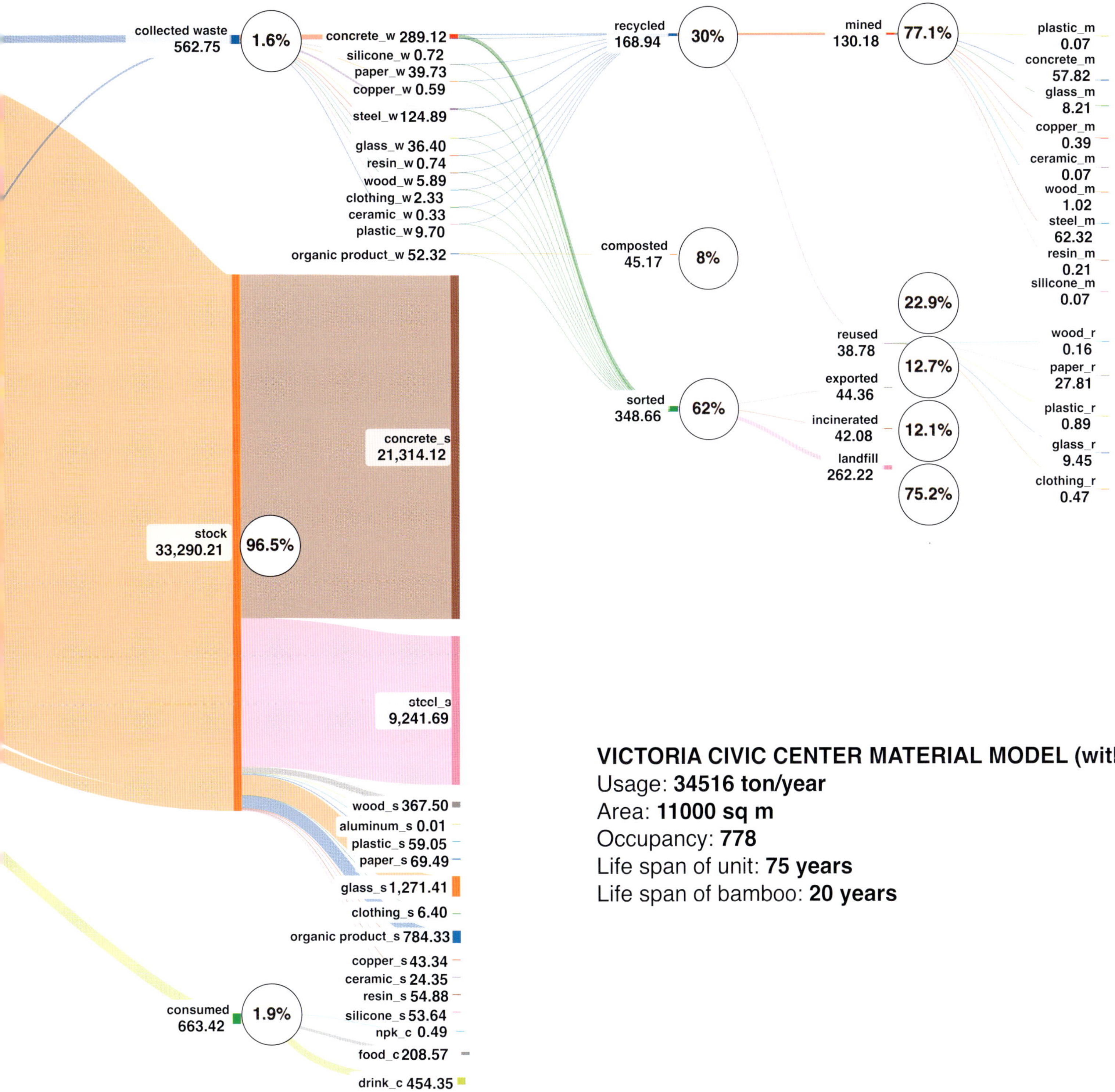

VICTORIA CIVIC CENTER MATERIAL MODEL (with bamboo facade details)

Usage: **34516 ton/year**

Area: **11000 sq m**

Occupancy: **778**

Life span of unit: **75 years**

Life span of bamboo: **20 years**

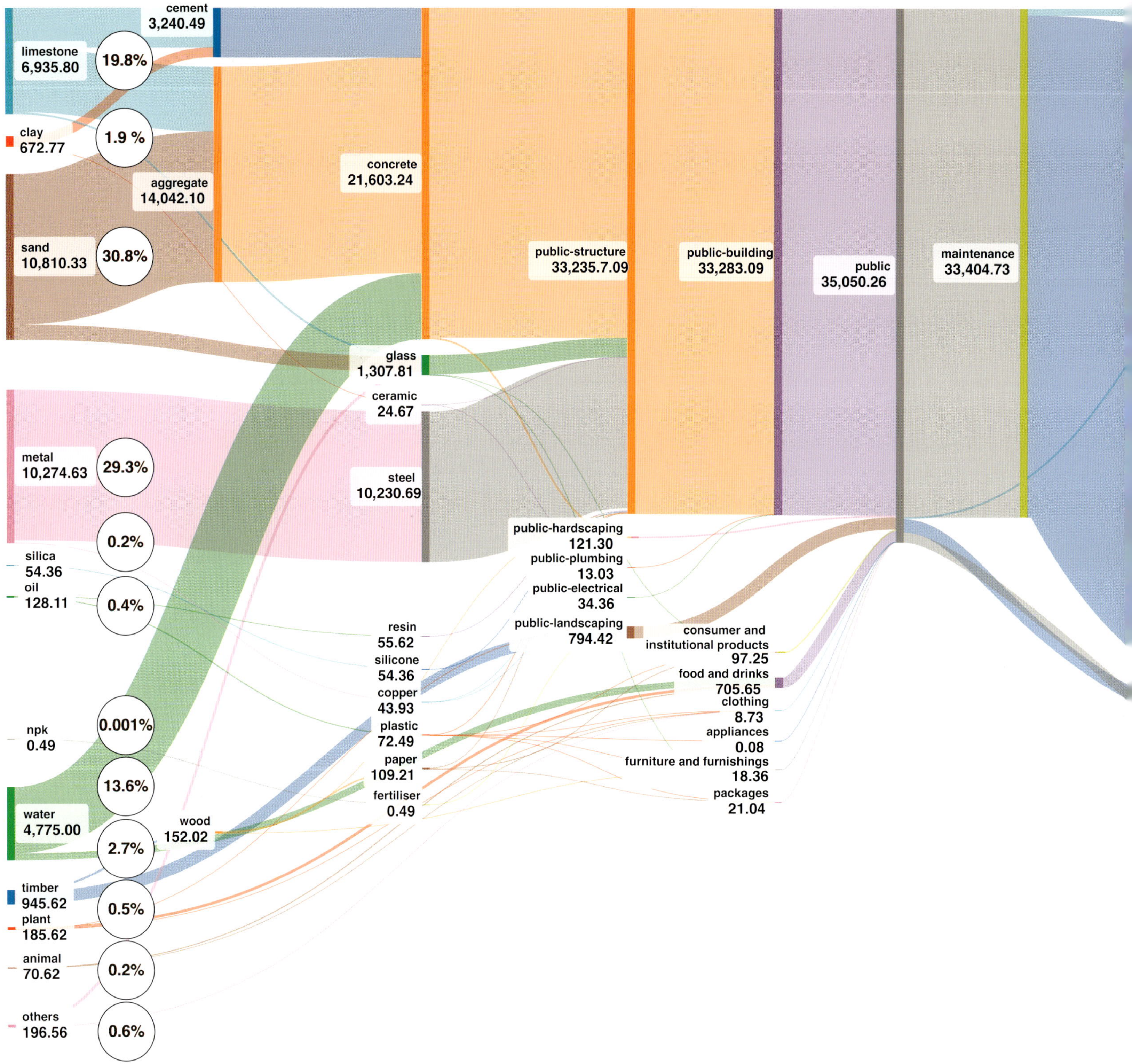
cement
3,240.49
limestone
6,935.80
19.8%
clay
672.77
1.9 %
aggregate
14,042.10
sand
10,810.33
30.8%
concrete
21,603.24
public-structure
33,235.7.09
public-building
33,283.09
public
35,050.26
maintenance
33,404.73
glass
1,307.81
ceramic
24.67
metal
10,274.63
29.3%
steel
10,230.69
0.2%
silica
54.36
oil
128.11
0.4%
public-hardscaping
121.30
public-plumbing
13.03
public-electrical
34.36
public-landscaping
794.42
resin
55.62
silicone
54.36
copper
43.93
plastic
72.49
paper
109.21
fertiliser
0.49
npk
0.49
0.001%
13.6%
water
4,775.00
2.7%
wood
152.02
timber
945.62
plant
185.62
0.5%
animal
70.62
0.2%
others
196.56
0.6%
consumer and institutional products
97.25
food and drinks
705.65
clothing
8.73
appliances
0.08
furniture and furnishings
18.36
packages
21.04

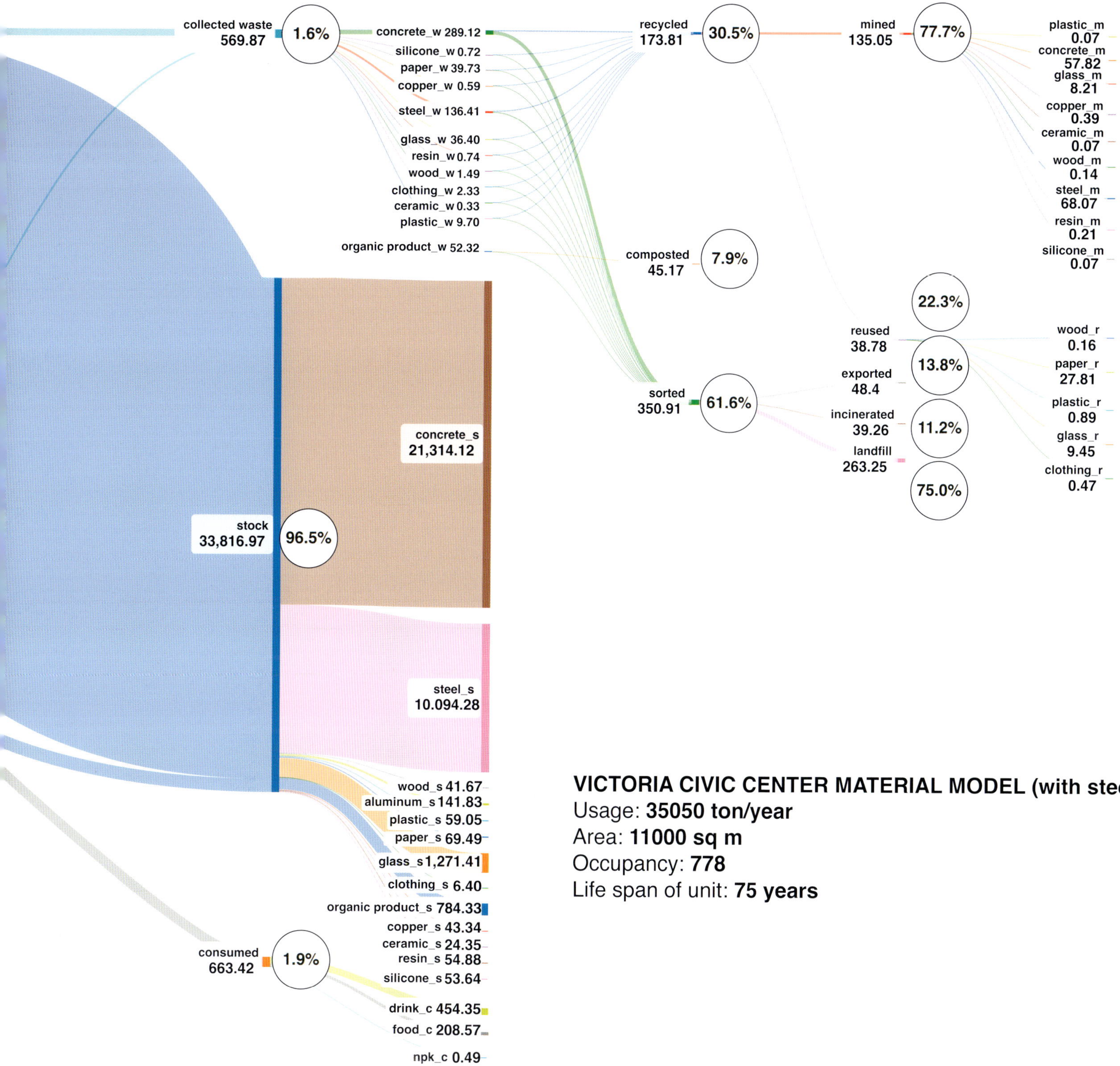

VICTORIA CIVIC CENTER MATERIAL MODEL (with steel facade and railing details)

Usage: **35050 ton/year**

Area: **11000 sq m**

Occupancy: **778**

Life span of unit: **75 years**

Situated within this garden is a chapel and event space built out of the very volcanic ash that covers the site. The architects and engineers tested a variety of volcanic ash mixtures to achieve the desired compressive strength for a series of customized interlocking bricks. The bricks are layered in a sandwich pattern to create an ultra-thin 200 mm diaphragm that can be shaped into vaults, spheres, and domes. Volcanic ash is also used as the cementitious mortar which holds the bricks together.

When we first see the chapel's exterior, it resembles a cascading chain of waves from the inside and the spine of a caterpillar from the outside. The architecture consists of a string of brick vaults made from the volcanic ash. The building is therefore made from the same material that covers the garden and is the basis for new plant growth. The interior space bubbles up, creating the sensation that the air is moving upward and lifting the vaults into the place. One gets the impression that emptiness and solidity are liaising to accommodate new life after the cataclysm.

The German-born American artist Kiki Smith forged the figure of a woman in blue-colored bronze held up by waves pouring out of her, in 2023, entitled "Dark Water." The sculpture makes the body of the woman appear weightless, the water seems like arms, and reverses the flow of things. The piece makes me think of the aesthetic possibilities of metabolic thinking. We witness connections, pay tribute to their creative power, and then imagine how these processes can be forged to make a better world for us and those that live within it.

FIG. 78: Image of the La Salle Chapel

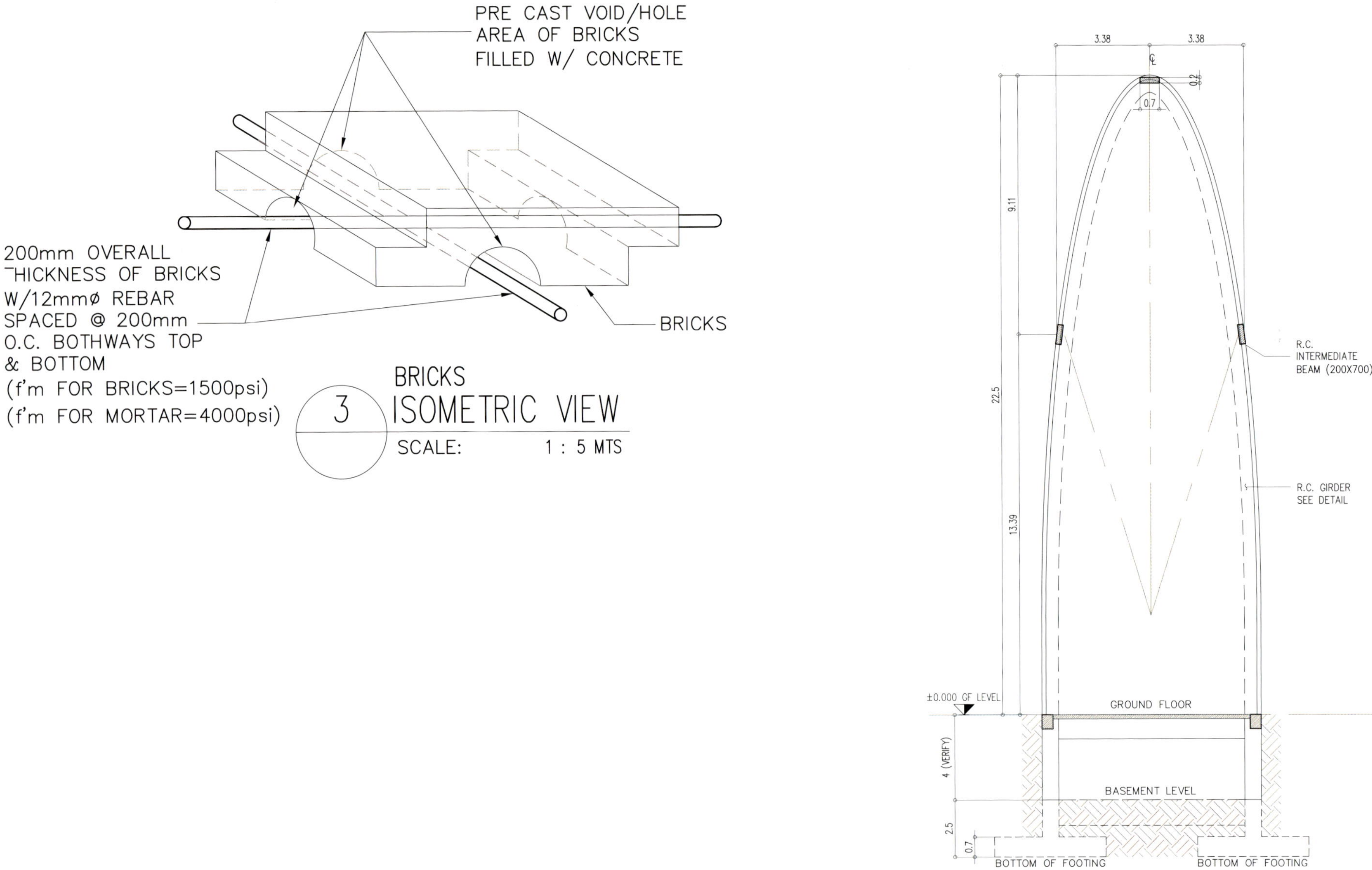

FIG. 79: Detail of the Interlocking Brick System.

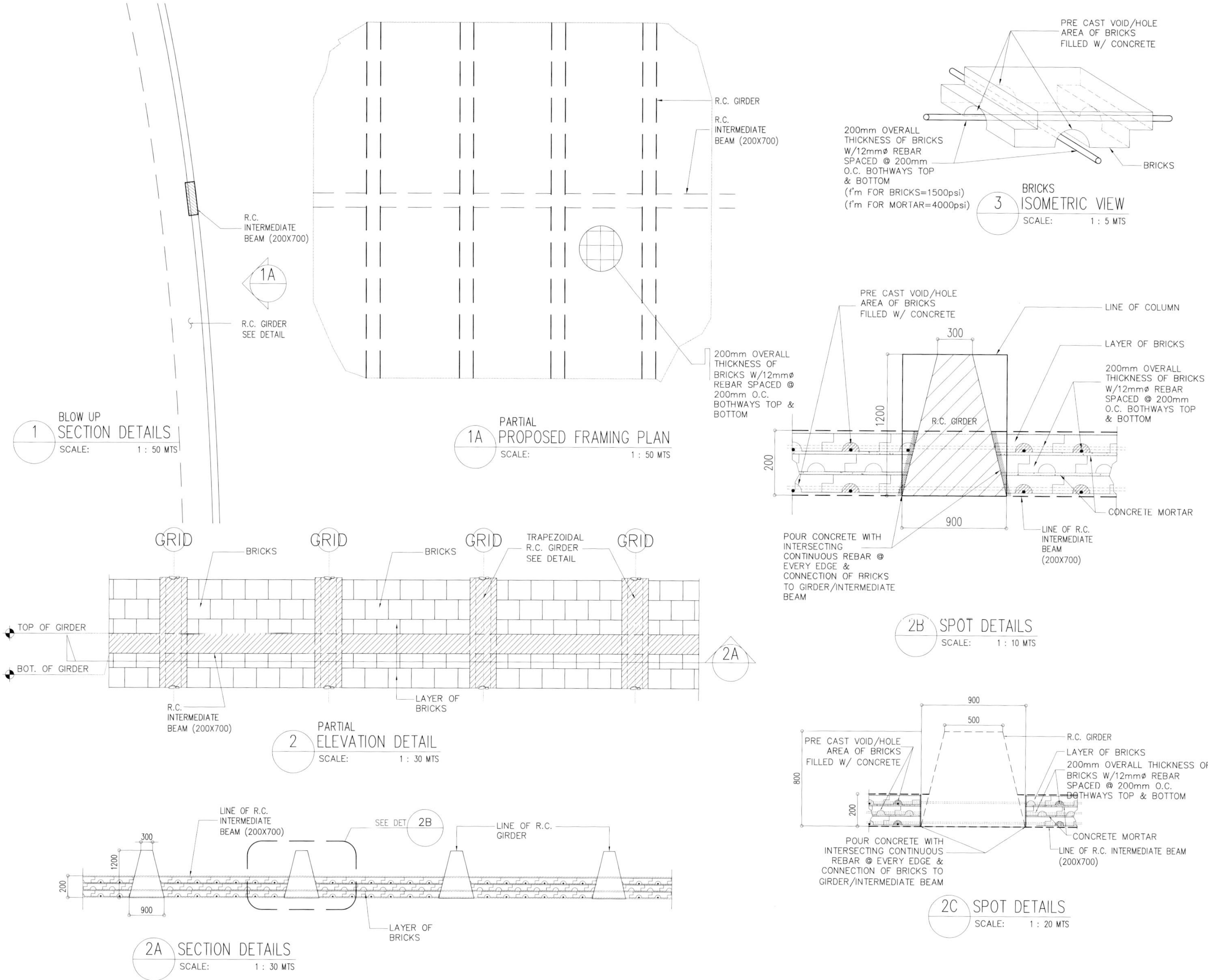
R.C. INTERMEDIATE BEAM (200X700)
R.C. GIRDER SEE DETAIL
1A
BLOW UP
1 SECTION DETAILS
SCALE: 1 : 50 MTS
R.C. GIRDER
R.C. INTERMEDIATE BEAM (200X700)
200mm OVERALL THICKNESS OF BRICKS W/12mmø REBAR SPACED @ 200mm O.C. BOTHWAYS TOP & BOTTOM
PARTIAL
1A PROPOSED FRAMING PLAN
SCALE: 1 : 50 MTS
PRE CAST VOID/HOLE AREA OF BRICKS FILLED W/ CONCRETE
200mm OVERALL THICKNESS OF BRICKS W/12mmø REBAR SPACED @ 200mm O.C. BOTHWAYS TOP & BOTTOM
(f'm FOR BRICKS=1500psi)
(f'm FOR MORTAR=4000psi)
BRICKS
BRICKS
3 ISOMETRIC VIEW
SCALE: 1 : 5 MTS
PRE CAST VOID/HOLE AREA OF BRICKS FILLED W/ CONCRETE
LINE OF COLUMN
LAYER OF BRICKS
200mm OVERALL THICKNESS OF BRICKS W/12mmø REBAR SPACED @ 200mm O.C. BOTHWAYS TOP & BOTTOM
300
1200
R.C. GIRDER
200
900
CONCRETE MORTAR
POUR CONCRETE WITH INTERSECTING CONTINUOUS REBAR @ EVERY EDGE & CONNECTION OF BRICKS TO GIRDER/INTERMEDIATE BEAM
LINE OF R.C. INTERMEDIATE BEAM (200X700)
2B SPOT DETAILS
SCALE: 1 : 10 MTS
GRID
GRID
GRID
GRID
BRICKS
BRICKS
TRAPEZOIDAL R.C. GIRDER SEE DETAIL
TOP OF GIRDER
BOT. OF GIRDER
2A
R.C. INTERMEDIATE BEAM (200X700)
LAYER OF BRICKS
PARTIAL
2 ELEVATION DETAIL
SCALE: 1 : 30 MTS
900
500
800
200
PRE CAST VOID/HOLE AREA OF BRICKS FILLED W/ CONCRETE
R.C. GIRDER
LAYER OF BRICKS
200mm OVERALL THICKNESS OF BRICKS W/12mmø REBAR SPACED @ 200mm O.C. BOTHWAYS TOP & BOTTOM
CONCRETE MORTAR
POUR CONCRETE WITH INTERSECTING CONTINUOUS REBAR @ EVERY EDGE & CONNECTION OF BRICKS TO GIRDER/INTERMEDIATE BEAM
LINE OF R.C. INTERMEDIATE BEAM (200X700)
2C SPOT DETAILS
SCALE: 1 : 20 MTS
LINE OF R.C. INTERMEDIATE BEAM (200X700)
SEE DET 2B
LINE OF R.C. GIRDER
300
1200
200
900
LAYER OF BRICKS
2A SECTION DETAILS
SCALE: 1 : 30 MTS

v. Modular Communal Infrastructure

Disaster-relief architecture is something of a chimera. Propelled by the noblest of intentions, funded by the widest global consortium of do-gooders, and underpinned by so many excellent design proposals, there should be no shortage of solutions. And yet, after every disaster, we hear the call for proposals and the gradual stalling of the bureaucratic wheels of fortune. Many successfully have been built, and yet countless others have failed because of a myriad of reasons.

Scavenging through the history of this building type, one common design denominator is clear: they all focus on the unit of inhabitation. Either the cot, the room, or the house, disaster relief is about the dwelling.

Yet, when we step back from the dwelling, we see a large collection of dwellings that make up a settlement. The settlements are planned with a barrack-style efficiency privileging density over any other concern.

Tamar Salinger from Pratt Institute researched a number of these disaster-relief settlements and found that they often did not consider the integration of waste removal, drainage, electrical hookup, and access to clean water in their planning logic. Her studies exposed the many ways infrastructure was missing from the design equation. The consequence of an obsession with the dwelling is the architectural blind spot of integrating the buildings with the many metabolic flows in the site.

Tamar developed a modular project whose design contribution was to rethink the communal infrastructures of disaster-relief settlements. She identified four critical metabolic flows, including water and electricity. She mapped how these might be distributed in a series of typical disaster-relief communities. Her architectural module is conceived as a connector between dwelling, site, and community. It is composed a recycled plastic polymer that is 3D printed into interlocking diagrid shapes.

The brilliance of the design is how these modular shapes create a structural lattice that brings clean water and electricity to the homes in a settlement while helping to drain out wastewater. Each metabolic flow has a different carrying capacity and thereby engineering requirement, creating a variation in the size-to-

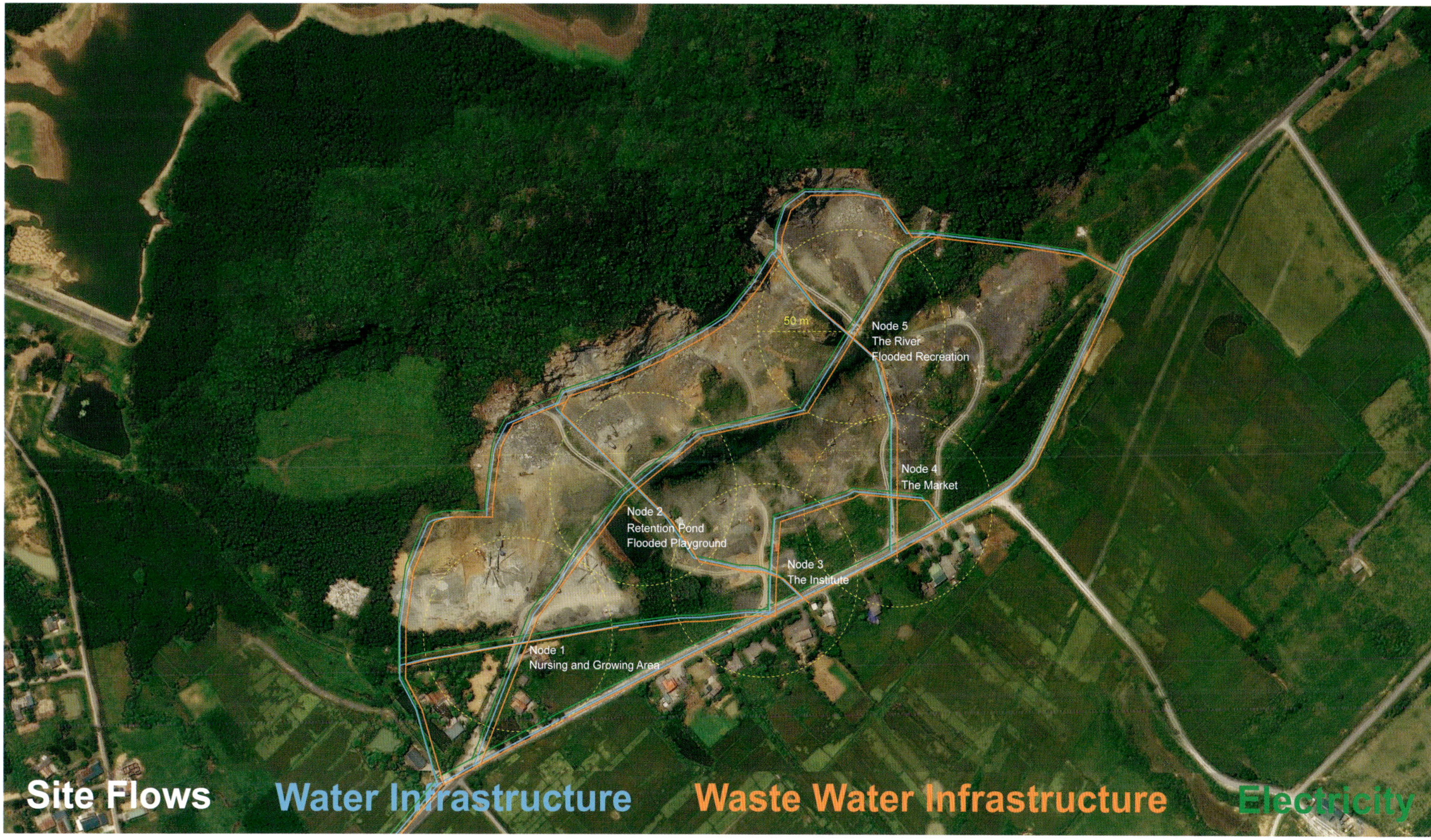

weight ratio of each modular connector. The different modular connectors result in different geometries for each lattice system.

One can imagine these systems being deployed in different geographies with their own set of community requirements. The result of which would be a diverse tapestry of infrastructural links endemic to each place and community - an architectural analog for the craft of weaving that is common throughout almost all regions, yet serves as a means for expression and identity for each community.

FIG. 80: Representation of the Modular Communal Infrastructure.

Integrated Infrastructure Systems

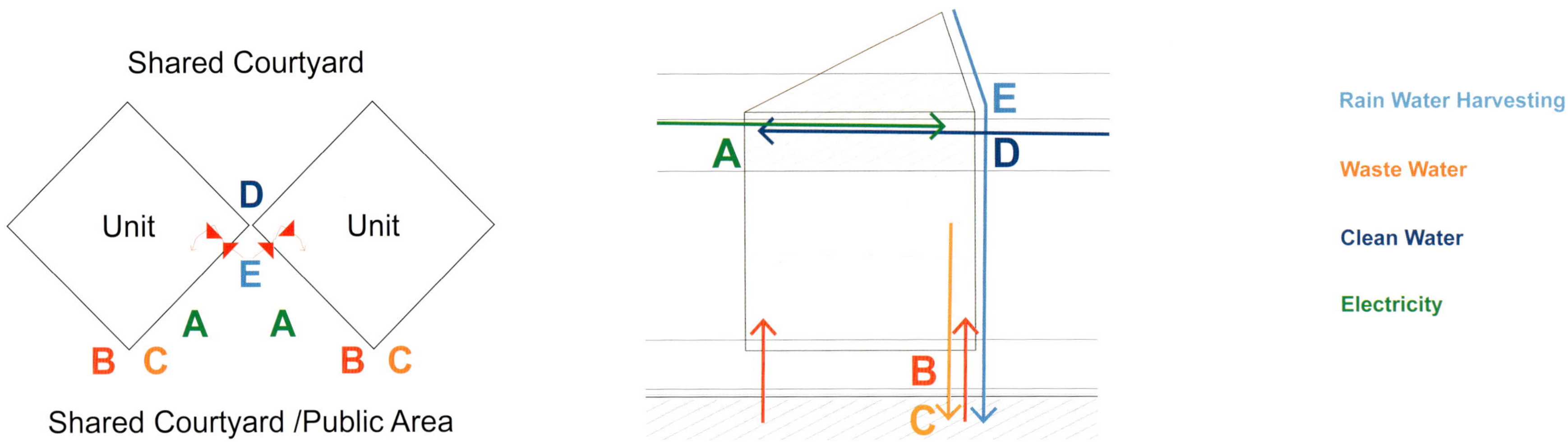

FIG.81: A Tapestry of Infrastructures.

Modular-metabolic Facade system

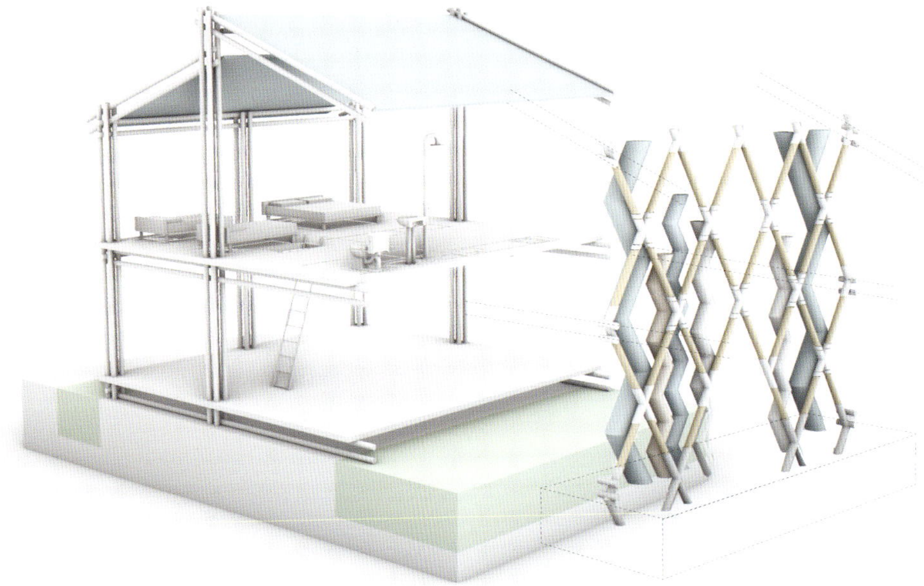

Rain Water Harvesting Waste Water Clean Water Electricity
XL L M S

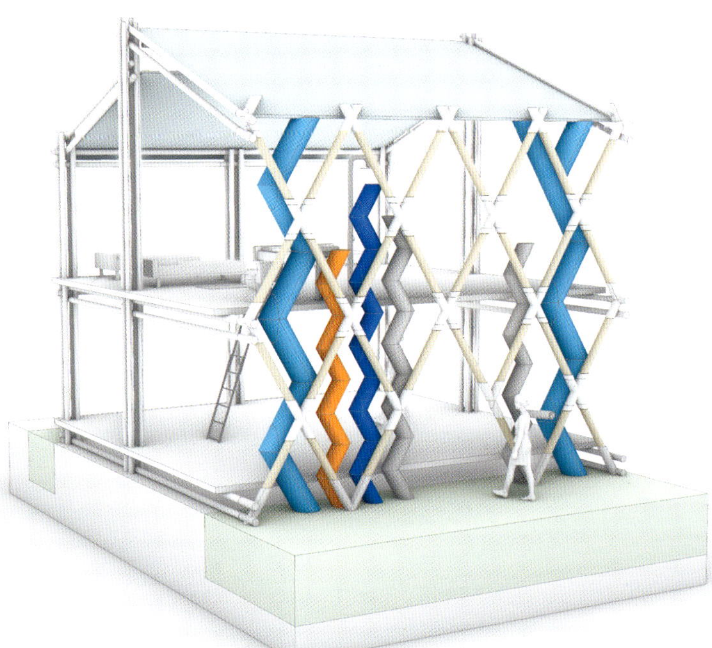

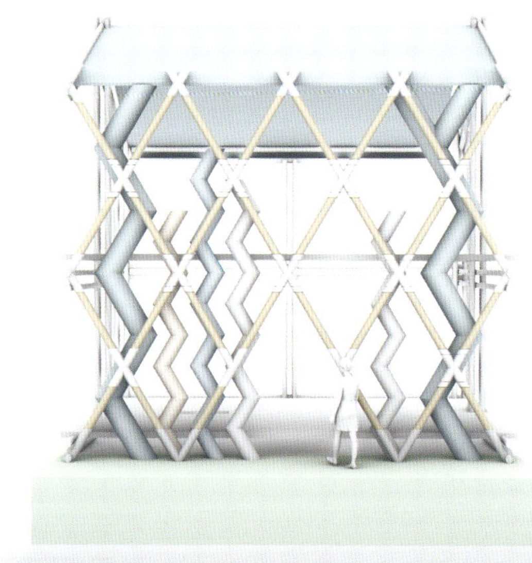

05

Data and Proxies

As discussed in Chapter 3, the collection and collation of data depicting aspects of the metabolic character of constructed environments is derived from numerous sources and is an important consideration, especially to the veracity of results of the stock-flow modeling. As characterized there, some seven broad categories of data are typically involved, ranging through spatial and population measures, construction data, information about particular stocks of materials, waste management, and so on. A primary source of this data is government agencies, though proprietary sources for particular kinds of materials are also available and relatively reliable. Also, the spatial resolution or specificity of data varies from directly observed and measured information to project-specific and locally applicable data, as well as to data pertaining to the area or region on which the metabolic assessment is being made. In what follows are some considerations about data quality and reliability, including the use of proxies beyond direct measurement and observation.

A. Proxy Measures.

As can be seen from Chapter 4 dealing with case studies, proxy measures are abundant and more the rule than the exception. Simply put, a proxy is a figure or estimate that can be used to represent a value of something in a calculation such as those involved in stock-flow modeling. Effectively it is a substitute for that aspect. Etymologically, proxy derives from the middle-English *procuracie* meaning to 'act on behalf' of something or someone. Usage here is also not to be confused with 'proxy servers' which in the digital world of computing allow data to be obtained from restricted sites via unrestricted or less restricted locations elsewhere. Like all manner of data, proxies are subject to issues of both veracity or accuracy of measurement and reliability covering the extent a proxy for something can be seen to obtain. If anything, there appears to be a growing trend towards more and better data about material, energy, and water consumption aspects of local urban developments. In part this is due to the adoption of more proactive environmental and social strategies both by local governmental authorities and private proprietary firms. Environmental accountability, for example, has been rising in noticeable ways since the mid-1990s if not earlier (Porter and van der Linde, 1995; and Aragan-Correa, Mathais-Reche, and Senise-Barrio, 2004). Social accountability, according to commentators, appears to have risen in strengthening degrees from interrelationships among stakeholders in three market drivers. They are the product market shaped by customers, the capital market shaped by investors, and the market of social pressure by governmental activists and non-governmental organizations concerned about improved social performance (Baron, Harjoto, and Jo, 2009). In particular, from this authoritative survey, corporate social performance was found to be increasingly responsive to the exercise of social pressures.

B. Sources of Data.

Reasonably available sources of data and certainly at the level and quality of government agencies conform largely to administrative units of governance, including transnational or international sources such as the World Bank and the United Nations. UNHabitat, for instance, is mandated by the United Nations General Assembly to promote socially and environmentally sustainable towns and cities by working with some 90 countries to provide transformative change in human settlements through knowledge, policy advice, technical assistance, and collaborative action. Perhaps less well known here are agencies like Statista and SEDAC. The former, Statista, is a global data and business intelligence platform with an extensive collection of statistics, reports, and insights in about 120 institutions, first established in Germany in 2007. It now operates in some 13 locations worldwide and employs about 1,100 professionals replete with interactive databases. SEDAC is the Socioeconomic Data Applications Center of NASA's Earth Observing System Data and Information System. It commonly provides georeferenced population data sets of nations, such as Mexico among the four cases depicted here, as well as for municipalities and other localities within Mexico.

Below the international level are relevant agencies within national administrative units, among the four cases of settlement coexistence discussed here. For example, France is comprised of 13 regions, including Corsica, 98 departments of local activity and with dependent subdivided units, such as *arrondisements*, roughly translated as districts, such as the 20 arrondisements of the Paris Commune. Each of these arrondisements is subdivided in IRIS, or *ILots Regroupes pour L'Information Statistique* (aggregated units for statistical information). Each IRIS includes between 1,800 and 5,000 inhabitants. The U.S. is subdivided from its formation in 1789 into 50 states, plus Washington D.C. and several other territories, alongside of counties and towns within states, like the town of Westwood in Norfolk County, Massachusetts. Mexico for its part also consists of states of which there are 31 plus Mexico City, all formed in 1917 and here including the state of Nuevo Leon and the municipality of Monterrey, within which there is the organic settlement of *La Campana*. China is comprised of 22 provinces administered by the central government, including here Guangdong Province which is further composed of 36 counties, 20 county-level cities, and 21 prefecture and sub-prefecture level cities. Thus, the village of *Chencun* is in the Shunde District of Foshan City, a prefecture level city in the province. Regarding data, most if not all these subnational-level units regularly furnish publicly available data, particularly about demography, aerial coverage information, and select environmental information.

More specifically, at the local level, certain reliable and continuous sources of data are also available. In Paris, for instance, there is the APUR or Atelier Parisien d'Urbanisme. It is a non-profit association of parties created in 1967 as a place for shared, forward-looking, multi-scalar study information. APUR documents analyze and imagine the urban and societal evolution of Paris, the 12 areas covered by the Greater Paris Public Territorial Establishment, and the Grand Paris Metropolis as a whole. It also contributes to defining public development policies,

97 Canton Street	
Location	97 Canton St
Acct#	00002981
Appraisal	$834,900
Building Count	1
Mblu	24/233///
Assessment	$834,900
PID	2981

Current Value (Appraisal)	
Valuation Year	2023
Improvements	$360,800
Land	$474,100
Total	$834,900

Current Value (Assessment)	
Valuation Year	2023
Improvements	$360,800
Land	$474,100
Total	$834,900

FIG.1: A Sample of Tax Assessment Data for Westwood, Massachusetts, U.S.A.

Building Photo

Building Information

Year Built	1994
Living Area	1,808
Replacement Cost	$434,731
Building Percent Good	83
Replacement Cost Less Depreciation	$360,800
Assessment	
PID	

Building Attributes

Style	Raised Ranch
Model	Residential
Grade	Average - Good
Stories	1
Occupancy	1
Exterior Wall 1	Vinyl Siding
Exterior Wall 2	
Roof Structure	Gable/Hip
Roof Cover	Asph/F Gls/Cmp
Interior Wall 1	Plastered

Building Attributes

Interior Wall 2	
Interior Floor 1	Hardwood
Interior Floor 2	
Heat Fuel	Gas
Heat Type	Forced Air-Duc
AC Type	Central
Total Bedrooms	3 Bedrooms
Total Bathrooms	2
Total Half Baths	0
Total Extra Fixtures	1
Total Rooms	6
Bath Style	Modern
Kitchen Style	Modern
Number of Kitchens	
Garage	
Fireplaces	1
Chimney Type	
Basement	

Building Layout

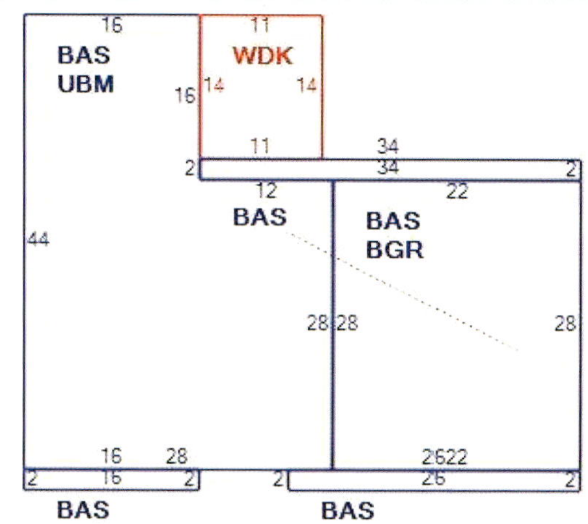

Building Sub-Areas (sq ft)

Code	Description	Gross Area	Living Area
BAS	**First Floor**	1,808	1,808
BGR	**Basement Garage**	616	0
UBM	**Basement, Unfinished**	1,040	0
WDK	**Deck, Wood**	154	0
		3,618	1,808

Land

Use Code	1010
Description	Single Fam MDL-01
Zone	RB
Neighborhood	C
Alt Land Appr	No
Category	
Size (Acres)	0.92
Frontage	
Depth	
Assessed Value	$474,100
Appraised Value	$474,100

Appraisal

Valuation Year	Improvements	Land	Total
2022	$376,250	$419,300	$795,500
2021	$387,300	$399,300	$788,600
2020	$345,800	$399,300	$745,100

Assessment

Valuation Year	Improvements	Land	Total
2022	$376,250	$419,300	$795,500
2021	$387,300	$399,300	$788,600
2020	$345,800	$399,300	$745,100

investigates and develops forward-looking studies, like the ZAC discussed here, and produces and places stress in a metropolitan culture. Its open data categories include reference data and data about facilities and services; economy, employment, and trade; habitat and housing; social and societal components; public space and mobility; history and heritage; as well as landscape and urban planning (APUR, 2010). A significant and consistent source of data for the building of Westwood in Massachusetts, U.S., is that provided for real estate tax assessment purposes under state law (Westwood, MA, 2023). As depicted in Figure 1 of this chapter, these data are comprehensive and precise, covering building information, including replacement and depreciation costs, building attributes including detailed accounting of materials used, as well as building layouts including plans and photographs, and all concluding with a monetary appraisal assessment. Moreover, these data are for every dwelling and building within the Westwood township of Norfolk County and by street address, making them geo-referenceable. Therefore, in this study, a relatively detailed and accurate database of buildings and building materials could be compiled. Specific data for La Campana in Monterrey, Mexico, was formed largely through available local government sources, as well as from field observations. For China use was made of available 'open data' as well as standardized data sets for built-up areas of the county (Liu, Song, et al., 2015; and Jiang, Sun, Guo, et al., 2021). Also as described in the following sections, the richness of local-level data concerning urban settlement is increasing with time as local authorities and other related agencies work more closely on matters of environmental performance of building and infrastructure development.

C. Emergent Issues.

Based primarily on work described here in Chapter 4 on case studies at various scales and locations, several issues emerged requiring attention in assessing data validity and reliability in metabolic stock-flow modeling of which five are particularly conspicuous. Perhaps first among these is the completeness of data ascribed to an element of the constructed environment and activities associated with it. What is meant by this is making sure that all relevant components of material, energy, and water flows are included. Some summary data bases, for instance about building construction data, focus on primary though not all building materials such as wood, concrete, and brick but exclude data about roofing materials and interior finishes (Center for Sustainable Systems, 2002). Within the research intentions of such databases this may be understandable and also be aimed to reflect major contributions, but when placed into the context of associated multiplier effects of rarer materials it can be misleading. Related is also the before mentioned spatial resolution of the data. Second, there is the role of the metabolic account of material flows in the projection of outcome measures. Per force of emphasis on climate change, for instance, embodied energy and carbon content are often important outcome measures. To obtain them requires expression of material flows in volumes or material weights, like metric tons per year, converted using embodied energy coefficients by material. Concrete, for example, has a relatively low coefficient of around 1.0 to 2.0 MJ/kg. Similarly, limestone is even less, whereas aluminum and even finishes like paint are much higher at around 150–200 MJ/kg and 90 MJ/kg, respectively. Therefore, a situation arises where a small amount of material can become the proxy for a large amount of a derivative like embodied energy. Hearsay accounts of concrete being so wasteful of embodied energy often also hinge on a confusion between sheer large volumes of material in use rather than precise low coefficients of embodied energy. Thirdly, care must be taken with aspects like fuel economy data in transportation to be sure that they reflect vehicle fleet characteristics as well as driving behavior. For the case studies in Chapter 4, these measures ranged from 1.2 KWh/km and above for American Peripheral developments reflecting large relatively inefficient vehicles in the mix and relatively cheap fuel, through 1.0 for informal development in Mexico with greater sensitivity to fuel price even with older vehicle fleets, down to 0.7 in Paris with both vehicle efficiencies and driver behavior manifesting lower values. In other words, a range is yielded, but one that is explicable. Broader ranges would suggest inaccuracies in the data being used. In short, inspection of ranges across data for the same material use can assist in assessing accuracy. Fourth, a detailed description of how data is compiled is important. For instance, data from Geographic Information Systems (GIS) regarding scales of building can be misleading by reflecting the cover footprint but not the actual built footprint. In the peripheral case of Westwood in the Boston area, for example, the GIS-based government data source seems to be roughly half the actual area of residential building compared to, say, Zillow proprietary real-estate listings (Mass.gov, 2019). Fifth and finally, the temporal frequency of data often varies and is beyond occurring annually. As the modeling of metabolism is temporal by nature, some data changes and adaptions become necessary.

D. Accountability.

Throughout the case studies pursued in this volume, all data describing material, energy, and water stock flows are accurately and comprehensibly cited and as depicted in Figure 2. There, a standard format consists of the stock type category like 'buildings,' any formula or particular coefficient numbers involved, comments about units of measure like x square meters, and the source of the information, including references to how calculations may have been made from general data categories like GIS sector layers. Therefore, the results are completely reproducible, even as different data may become available or appear. Indeed, this aspect of accountability and reproducibility of results is an important feature of the stock-flow modeling process as is the sheer conservation of matter throughout this process. Also, representation in the form of Sankey diagrams is also summarized by ascribing values to different stages in the stock-flow. These range from material extraction and processing to end uses, overall stock consumption, and waste management, as depicted throughout Chapter 4. What remains are summary data depicting, for example, which domain of coexistence is the least harmful environmentally. Weight per time of material in the model gives an assessment across domain categories as to the sheer amount of material needed from initial resource domains. However, given that small quantities by weight of some manufactured materials involve sizable amounts of embodied characteristics harmful to the environment, other overall measures may be required for assessment. In essence stock-flow models of this kind are a relational sequence of potential proxies or multiple measures.

FIG. 2: A Sample of the General Format for Citing Data About Material Flows.

Material Data Source Westwood/Norwood

data	formula/number	comments	source
Buildings (m2)			
Total Building area	5832042.1	m2	1.012006686
Commercial Buildings area	1964837.5	m2	0.336903861
Residential Buildings area	3676004.4	m2	0.630311705
Manufacturing Buildings area	261223.7	m2	0.04479112
Roads (m)			
Length (m)	683,123.10	estimate width??	calculated from GIS vector layers
Commercial + Manufacturing (m2)		based on USA commercial percentages breakdown	https://www.eia.gov/consumption/commercial/data/2018/pdf/CBECS_2018_Building_Characteristics_Flipbook.pdf
Total	2226061.2		https://www.eia.gov/consumption/commercial/data/2018/pdf/CBECS_2018_Building_Characteristics_Flipbook.pdf
Stores	267127.344	mercantile 12%	https://www.eia.gov/consumption/commercial/data/2018/pdf/CBECS_2018_Building_Characteristics_Flipbook.pdf
Industrial	400691.016	(warehouse and storage + mercantile) 18%	https://www.eia.gov/consumption/commercial/data/2018/pdf/CBECS_2018_Building_Characteristics_Flipbook.pdf
Offices	690078.972	(offices + services + lodging) = 17%+7%+7% = 31%	https://www.eia.gov/consumption/commercial/data/2018/pdf/CBECS_2018_Building_Characteristics_Flipbook.pdf
Schools	311648.568	(education) 14%	https://www.eia.gov/consumption/commercial/data/2018/pdf/CBECS_2018_Building_Characteristics_Flipbook.pdf
Health	267127.344	(check - this is lumped with 'other' + food services - might be too high?) 12%	https://www.eia.gov/consumption/commercial/data/2018/pdf/CBECS_2018_Building_Characteristics_Flipbook.pdf
Public	289387.956	(public assembly + religious worship) 13%	https://www.eia.gov/consumption/commercial/data/2018/pdf/CBECS_2018_Building_Characteristics_Flipbook.pdf

TABLE 2: PRIMARY STRUCTURAL MATERIAL OF NEW LOW-RISE NONRESIDENTIAL BUILDINGS
BY BUILDING TYPE, UNITED STATES 2011

USDA: Wood and Other Materials Used to Construct Nonresidential Buildings in the United States: https://www.fs.usda.gov/research/treesearch/43783

	Construction Value		New Buildings & Additions				New Buildings & Additions			
	All Mil $	New & Additions Mil $	Wood Th ft²	Concrete Th ft²	Steel Th ft²	TOTAL Th ft²	Wood %	Concrete %	Steel %	TOTAL %
Stores	$19,796	$9,889	10,708	30,952	110,794	152,454	7.0%	20.3%	72.7%	100.0%
Industrial	$9,377	$7,203	6,416	39,960	39,458	85,833	7.5%	46.6%	46.0%	100.0%
Offices	$12,294	$9,236	10,715	19,864	56,034	86,613	12.4%	22.9%	64.7%	100.0%
Hotels	$3,542	$2,858	11,499	3,695	5,973	21,167	54.3%	17.5%	28.2%	100.0%
Schools	$47,511	$19,050	13,115	34,877	104,054	152,045	8.6%	22.9%	68.4%	100.0%
Colleges	$23,547	$12,854	9,249	11,879	30,931	52,060	17.8%	22.8%	59.4%	100.0%
Religious	$2,709	$1,636	4,443	2,521	6,198	13,162	33.8%	19.2%	47.1%	100.0%
Health	$31,465	$15,820	12,432	12,263	45,750	70,446	17.6%	17.4%	64.9%	100.0%
Public	$47,302	$29,588	5,182	16,806	49,718	71,705	7.2%	23.4%	69.3%	100.0%
Recreation	$4,584	$1,372	2,369	4,845	11,579	18,793	12.6%	25.8%	61.6%	100.0%
Misc	$6,251	$3,394	2,543	3,091	3,198	8,832	28.8%	35.0%	36.2%	100.0%
All Buildings	$208,380	$112,900	88,672	180,752	463,686	733,110	12.1%	24.7%	63.2%	100.0%

Table 2: Primary Structural Material by Building Type

data	formula/number	comments	source
weight of 1 cubic meter wood	870	equal average of hardwood and softwood, in kilograms	https://extension.psu.edu/calculating-the-green-weight-of-wood-species#:~:text=With%20the%20specific%20gravity%20around,3)%2C%20at%20theoretical%20most%E2%80%A6
1 board foot equal cubic meter	0.0023597		
wood weight for store for us	313716979	total board foot equivalent * the equivalent of one board foot in cubic meter * weight of one cubic meter wood / data for the US 2011	https://www.fpl.fs.fed.us/documnts/pdf2013/fpl_2013_adair001.pdf
wood weight for industrial for us	9622288	total board foot equivalent * the equivalent of one board foot in cubic meter * weight of one cubic meter wood / data for the US 2011	https://www.fpl.fs.fed.us/documnts/pdf2013/fpl_2013_adair001.pdf
wood weight for office for us	236145364	total board foot equivalent * the equivalent of one board foot in cubic meter * weight of one cubic meter wood / data for the US 2011	https://www.fpl.fs.fed.us/documnts/pdf2013/fpl_2013_adair001.pdf
wood weight for school for us	384848416	total board foot equivalent * the equivalent of one board foot in cubic meter * weight of one cubic meter wood / data for the US 2011	https://www.fpl.fs.fed.us/documnts/pdf2013/fpl_2013_adair001.pdf
wood weight for health for us	469053189	total board foot equivalent * the equivalent of one board foot in cubic meter * weight of one cubic meter wood / data for the US 2011	https://www.fpl.fs.fed.us/documnts/pdf2013/fpl_2013_adair001.pdf
wood weight for public for us	205939961	total board foot equivalent * the equivalent of one board foot in cubic meter * weight of one cubic meter wood / data for the US 2011	https://www.fpl.fs.fed.us/documnts/pdf2013/fpl_2013_adair001.pdf
total number of square meters for stores in the US	10921348790	117556.42 million sq feet	https://www-statista-com.ezp-prod1.hul.harvard.edu/statistics/1202914/total-square-feet-retail-properties-united-states/

06

Aspects of Improved Performance

Stock-flow accounts of the kind presented in this volume present depictions of various kinds of material flows. As almost literal rewritings of metabolic character and performance they are useful in and of themselves. Nevertheless, they also leave open matters of which measures are most useful and how this modeling can address matters of risk, resilience, and obsolescence of constructed environments and buildings. In what follows these topics are taken up along with discussion of so-called 'cradle-to-grave' and far more radical attempts as 'cradle-to-cradle' approaches to material stocks and flows within built environments.

A. Summary Measures.

Looking across the four case studies of settlement coexistence, several summary measures of relative metabolic performance appear possible. With regard to material flows as well as others, for instance, what might be termed 'user measures' can be applied, such as tons of material per year per population or amount of collected waste per population. In these regards as shown in Figure 1, the relative order of performance is led by La Campana with only 36 metric tons and 1.91 metric tons per population per year of total material use and collected waste; followed by the Clichy-Batignolles ZAC at 117 and 2.17 metric tons per year per population, respectively; and then Chencun at 117 and 7.33 metric tons per population per year; and tailed off by Westwood at 247 and 9.31 metric tons per year per population for overall production and collected waste. As stated elsewhere these numbers reflect the overall high consumption and wastefulness

	Materials Tns/yr/pop	Collected Waste Tns/yr/pop	Ave. Embodied Energy MJ/tns/yr
Paris ZAC	117	2.17	3.2
Westwood	247	9.31	5.8
La Campana	36	1.91	2.8
Chencun	117	7.33	4.0

FIG. 1: Selected User Measures by Case.

of American peripheral development and the paucity of informal settlements. The comparative advantage of user-based metrics is that they relate directly to behavior in metabolic processes of settlement and, therefore, presumably to user-oriented policy and regulation. However, a case can also be made in the form of object-oriented ontologies and, therefore, material measures of things in themselves, or what is the performance of the built and constructed environment intrinsically. Here, the average measure of embodied energy and embodied carbon per year is probably best. In this regard the relative order of performance remains the same, as shown in Figure 1, with La Campana at 2.8 MJ/year, followed by the Clichy-Batignolles ZAC at 3.2 MJ/year, Chencun at 4.0 MJ/year, and Westwood at 5.8 MJ/year. Admittedly in the modeling these measures do account for behavior as the settlements were occupied, but also mainly reflect the material aspect of the way they were built and, therefore, a useful approximation of intrinsic performance.

With regard to other operational features of settlement coexistence, water conservation can and does occur through treatment and recycling. For instance, in the most consumptive 'Dispersed' model at Westwood on the U.S., some 58 percent of wastewater could be recycled from the treatment plant, also resulting in an 80 percent reduction in ground and surface water initially entering the system. In addition, technical substitution in sewage treatment in the form of membrane technologies, microfiltration, and so on can produce so-called 'New Water' that is potable and drinkable, although with higher use of electrical energy. In short, for water stressed areas, such as those in China, for instance, combinations of recycling and treatment can produce significant savings in source volumes of water, including water independence as in the case of Singapore (Rowe and Hee, 2019). For energy, prominent indicators of improved performance come by way of 'consumed' versus 'lost' proportions at the end of the stock-flow cycle. Again, in the highly consumptive case of Westwood peripheral development, the ratio of consumed to lost under existing conditions of 40:60 percent was reversed, representing a net 50 percent improvement. There and elsewhere other measures of superior performance came by way of reductions in the proportions of fossil fuels used at the source of energy models and the use of on-site heating and cooling technologies, resulting in a shift from 8 percent use of renewables at source to an equivalent close to 44 percent. In short, additional overall performance measures related to water recycling and treatment, alongside consumption and renewable sources are produced in these exercises as useful guides.

B. Sustainability, Risk, and Resilience.

Sustainability is a well-recognized feature of settlement coexistence for reasonably obvious reasons and can be assessed in the context of risks and resilience associated with constructed environments. Risk assessment often used in conjunction with sustainability usually refers to possible mishaps and likely tolerances associated with them. Resilience usually refers to the flexibility of settlement coexistences in resisting risks and their ability to adapt to sustainable directions or strategies. In the context of the modeling being described and discussed here, this can be engaged through cycles of repair, renovation, and reuse of existing structures.

i. Risk Assessment.

According to prominent figures like Benoit Mandelbrot (1924–2010), the Polish-born French American mathematician and polymath, risk may be categorized as being either 'mild' or

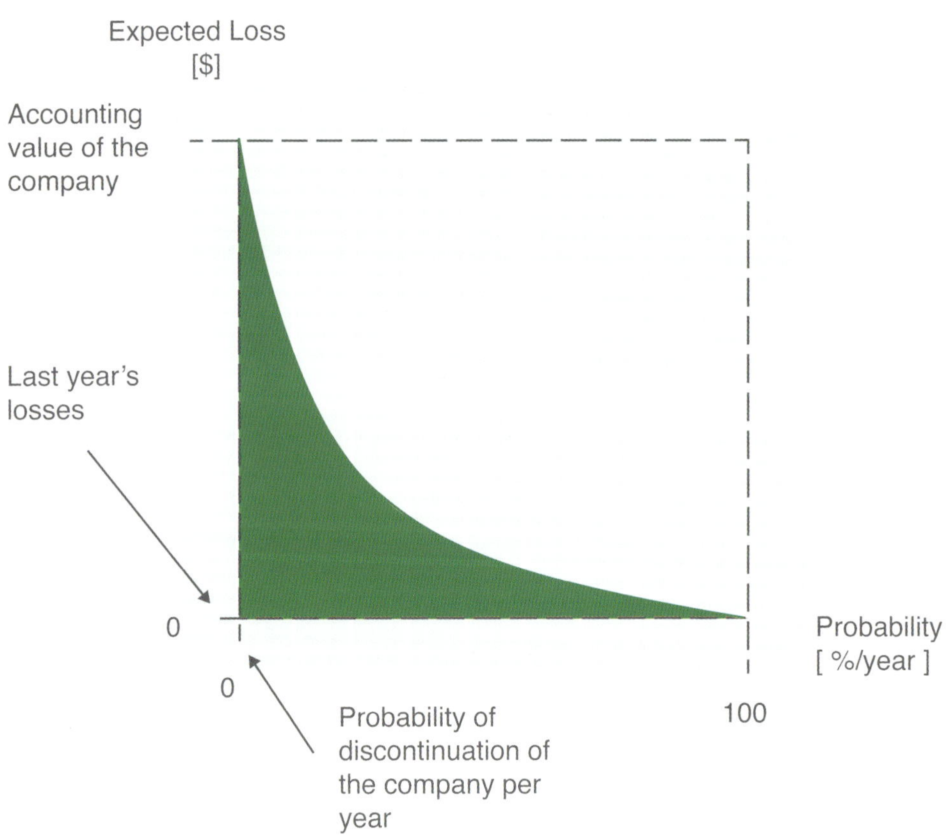

FIG. 2: Risk Assessment from Financial Point of View.

'wild,' where the former term applies to normal or near-normal probability distributions of happening. 'Wild' risk, by contrast, follows fat-tailed distributions in mathematical form and is far harder and often impossible to predict. (Mandelbrot and Hudson, 2008). From a financial point of view risk assessment can be described by Figure 2 and by a non-linear expression of expected loss to the probability of that occurrence and also with the continuation or discontinuation of the firm involved in any given year. In effect it is dynamic and different metrics apply (Popov, Lyon, and Hollcroft, 2016). For instance, often in public health and environment, expressions are made relating to changes in states usually in terms of risks to populations. Clearly related to settlement coexistence, for example, are conditions of urban greenery and biodiversity, carbon sequestration, and green building management, aspects that follow related to assumed benefits of thermal comfort, improved air quality, improved energy performance, and the like. Within biodiversity measures of 'endism' or extinction of species are routinely made often attributable to deforestation and urban development. Other related measures such as carbon sequestration also exist with mangroves rating very highly, for instance, and temperate forests less so. Also somewhat counterintuitively, forestation *per se* can be confounded by the biogenic component in tropical areas, such as Singapore, making them measurably net exporters and not sequesters of carbon (Rowe and Hee, 2019). More generally, risk assessment regarding human settlement focuses, nowadays anyway, on the consequences of climate change and on disaster risk reduction. In this latter regard, the *Sendai Framework for Disaster Risk Reduction, 2015-2030*, was adopted by the United Nations in Japan, strengthening disaster risk governance and management, particularly with regard to global mortality, economic loss, early warning systems, and climate change adaption (United Nations, 2015).

ii. Resilience.

Simply put, resilience is the capacity to withstand or recover from difficulties such as those imposed by risks. As goes risk so goes resilience, as it were. It is a kind of toughness and adaptability by stakeholders to recover from disasters and impending risks.

Largely it is often a form of coping behavior. Within the domains of coexistence under discussion here it also largely concerns physical aspects of constructed environments and material flows. By and large, here there are at least three kinds of resistive or resilient practices. First, there is continued maintenance and refurbishment of existing facilities. Second, there is involvement with adaptive reuse and recycling of facilities, and third, there is involvement with technical substitution of aspects of material flows with more efficient procedures, such as on-site solar arrays for sources of power. Repair and refurbishing largely involve routine maintenance though of increasing intensity. However, fundamentally buildings or components of constructed environments remain as they are, albeit in an upgraded state but with little material change. In principle, adaptive reuse and recycling depends on the reciprocal relationship of the old, constructed environment and the new environment typically accommodating different functions. Moreover, this may occur in at least three different manners, such as adding on through a change of interior function; addition, transformation, and conversion of the environment; and coexistence, imposition, and fusion with the environment. In other words, a building, for instance, is maintained and its internal function is modified, a building is modified to make it more efficient including modification of interior functions, and, finally, a host building is strongly modified and fused with an adjacent new structure. Similar principles might also apply to the infrastructure of streets and roads. This might involve changing the paving, diminishing the lanes of motorized traffic and increasing layers of bicycle and pedestrian lanes as at the Champs-Élyséess, or replacing paving with porous vegetated material all the way up to greenways and parkways. Technical substitution can also take on several forms, most of which have already been identified in the earlier case studies such as those at the Clichy-Batignolles ZAC and the use of solar arrays and on-site heat pumps for heating and cooling at Westwood. In addition, redesign of buildings and infrastructure may also be done, again like those at Westwood in the form of ecological corridors running through the site. Clearly, there are economic costs and benefits as well as different groups of stakeholders to also be considered. Moreover, they appear likely to mirror those involved in risk assessment briefly discussed earlier.

C. Sustainability and Obsolescence.

Looking more closely at the concept of obsolescence in relation to sustainability, beyond the manner it was described earlier in Chapter 3, is useful in addition to consideration of waste management and notions of end-of-life spans of structures. Clearly and as suggested, earlier interest in obsolescence rose during the latter part of the 20th century towards the reuse of existing stock away from excessive dominance of new buildings. Like machines and durable goods, buildings were to be eliminated and replaced when they became obsolete. Both rises of interests in historic conservation and environmental sustainability gave impetus to minimizing obsolescence and extending the longevity of buildings and constructed environments as physical, economic, social, and cultural assets. Also, blatant policies of 'planned obsolescence' were frequently frowned upon, as a moral stance emerged embracing longer-term considerations and reuse of stocks of various kinds (Packard, 1957).

i. Obsolescence

At root obsolescence refers to something no longer useful, old-fashioned, and vestigial. As such it is not a natural phenomenon but one that is a matter of human action and usually associated with artifices of one sort or another. Also, as such it is a cause rather than an end in itself, leading often to potential demolition. Causes of obsolescence can be manifold—physical, economic, functional, technical, social, legal, and political. Physically, for instance, materials in artifacts deteriorate. Economically, a type of building, for instance, and its location can be meaningful besides its quality. Also, the cost of continuing use compared to the expense of replacing it must be considered. Functionally, flexibility in a building's use, for example, plays a role. Technically, components of buildings have different renewal cycles, like elevators for instance. Socially, market and other factors come into play, such as an area's reputation, style, historic significance, and related characteristics. Legally and politically, failure to comply with shifting standards, often defined by cultural factors, can quickly result in the obsolescence of nonconforming buildings.

FIG.3: Paths of Performance by Time.

Ultimately, obsolescence is primarily responsible for the conformance of a building or structure to acceptable levels of performance in all respects and over time. This, in turn, can be represented by a two-space of 'performance' against 'time' as shown in Figure 3. There the profile of a building's performance rises to some higher or optimal level, usually above some overall acceptance threshold, before declining more rapidly with no maintenance, for instance, other than regular operation. Further longevity can be brought about to a structure through repair and refurbishment, thus arresting depreciation at least for some period of time (Thomsen and van der Flier, 2011). For some structures, like the US Capitol in Washington, D.C., for instance, this cycle can be portrayed presumably indefinitely. However, even in the field of outright historic preservation, arguments do arise about whether and how to preserve structures, even in perpetuity. Theoretically, if the renewal and regeneration cycles

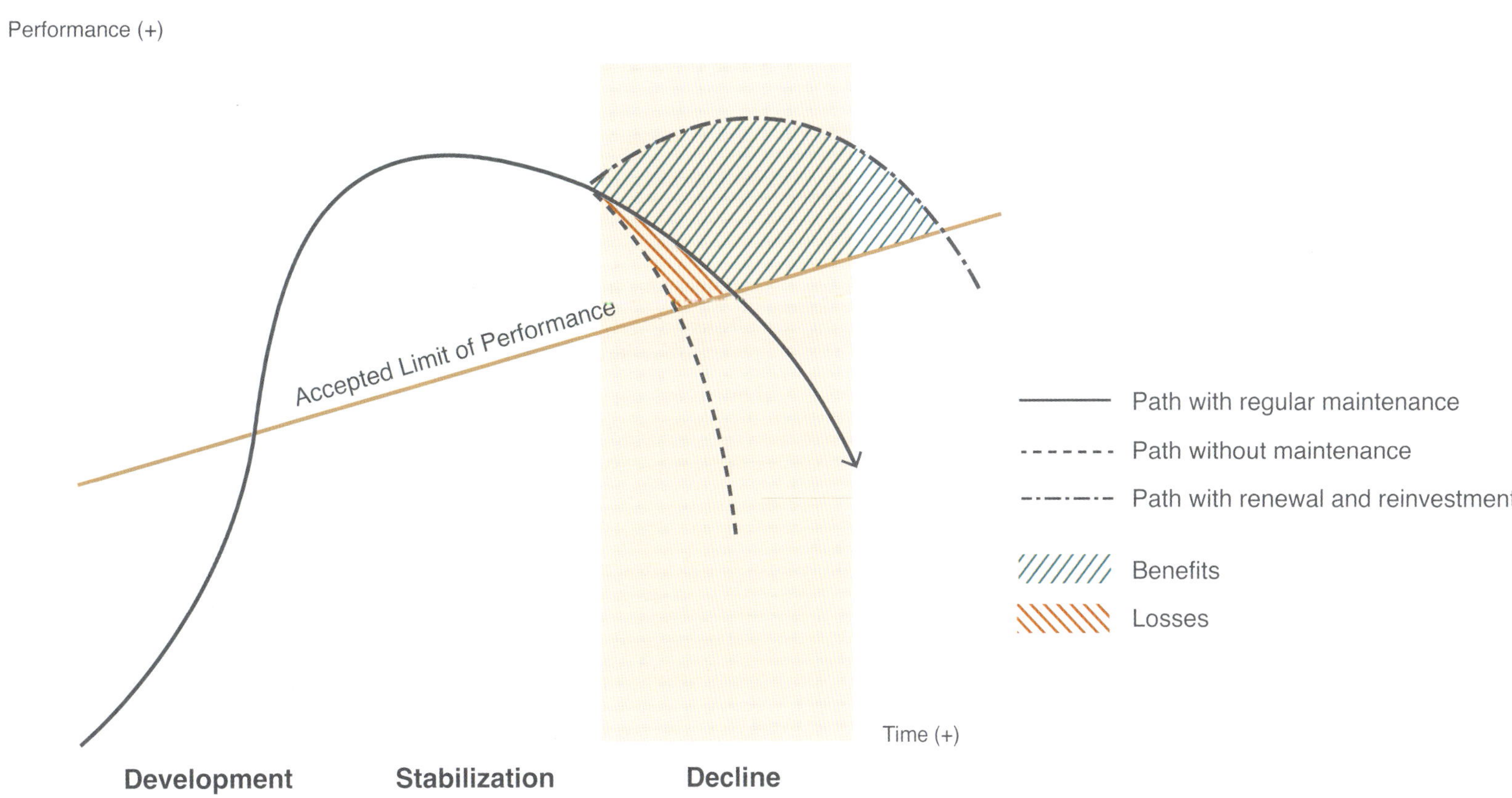

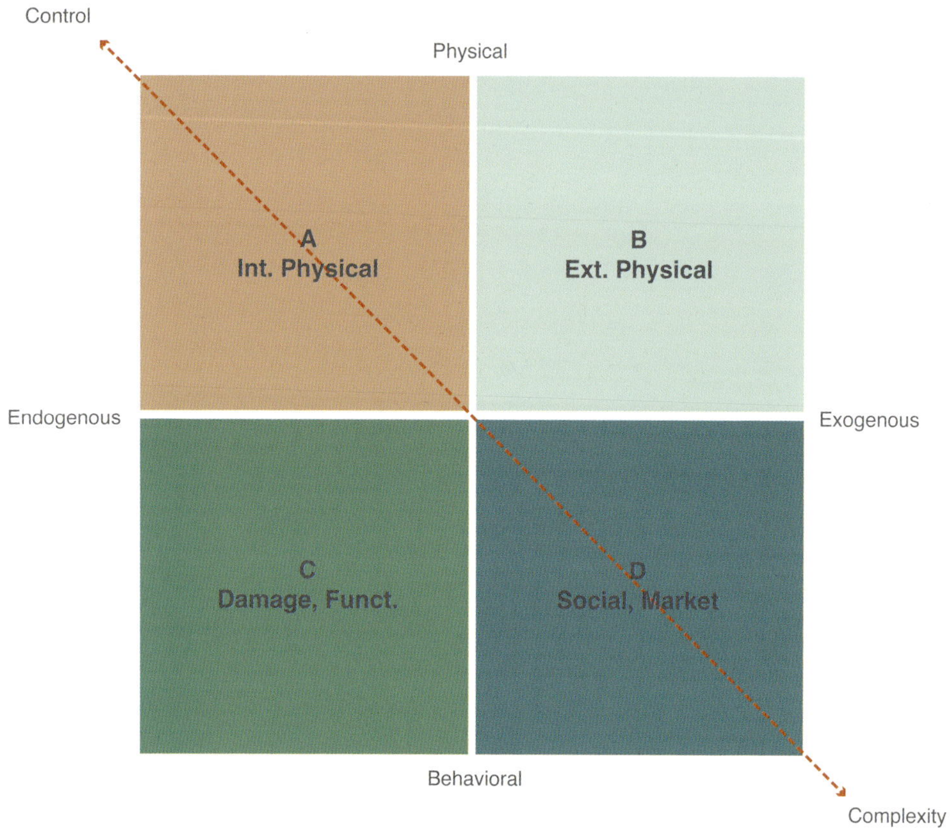

FIG.4: Dimensions of Obsolescence.

are extreme enough, the benefits deriving from the revived building will outweigh the losses or costs, as shown by the shaded areas in Figure 3, and consequently obsolescence, depreciation, and even demolition will also be prolonged.

In yet another framework, obsolescence can be seen to occur along a vertical axis with graduations between the intrinsic physical character of a structure through to its response to behavioral circumstances, like functional use. On another horizontal orthogonal axis, the variation is from endogenous occurrences conceived, say, with the future make up of a structure, through to exogenous circumstances affecting its life in the broader social and cultural context, all as shown in Figure 4. Clearly, the aspect of 'control' increases towards the upper left-hand corner of the two-space, whereas 'complexity' increases to the bottom right-hand corner. Moreover, it can be suggested that as a market becomes more competitive and mixed, product life spans will tend to increase. In other words, competing models tend to result in more sustainable ways rather than less (Rowe, 2017). As discussed previously with resilience, there are at least three kinds of practices for resisting obsolescence, ranging from continued maintenance and refurbishing, through adaptive reuse and recycling, to radical technical substitution into material flows. All told, they are aimed at some four aspects of obsolescence including longevity of buildings and constructed environments as such; being in an appropriate location at the right time given urban dynamics; basic infrastructural and supporting improvements; and through the manners of collected waste management.

Also, within this discussion often-preferred environmental technologies such as solar panels have relatively short and finite service lives. Photovoltaic panels, for instance, have life spans of from 25 to 30 years before being replaced. Also, during this span of time electrical generation deteriorates as well. Heat pumps, another often-preferred technology, last from 10 to 15 years and wind turbines last from 20 to 25 years before replacement (Walker, 2002; and Peterson, 2022). This differential of from 10 to 30 or so years also spills over into common elements of many residential structures, for example. Siding can be as low as 10 to 20 years for vinyl siding and up to 100 years or more for brickwork. Internal systems like electrical wiring vary with the period of construction, averaging around 70 years, whereas plumbing often depends on supply lines at 20 to 70 years or drainage lines at 70 to 100 years. Furthermore, the recycling of elements of environmental technology like solar panels is a mounting issue with some 80 million metric tons of waste globally, likely by 2050 (Walker, 2002). There the main aspect is the recovery of crystalline silicon that also requires relatively special recycling facilities. In addition, the use of rare earth elements is becoming more abundant, with China currently dominating the supply side. However, mines are often in countries like Myanmar where environmental and social standards are low, often leading to human rights violations, deforestation, and contamination of land and water (Ali, 2014). However, further socio-political analysis and consideration is beyond the scope of this present volume.

ii. Different Approaches.

Also as noted in Chapter 2 of this volume, as well as among the Chapter 4 case studies, the prospect of circular economies and cycles of recycling can and have had profound impacts on metabolism and the efficacy of improved environmental performance and sustainability. Among these, as noted by others, 'cradle-to-cradle' proposals for production by the likes of William McDonough and Michael Braungart—an architect and a chemist—have radical implications for even eliminating obsolescence in artifactual environments and consumer products. Their proposal is at once a political call to action by challenging the one-way 'cradle-to-grave' depicted by industrial production in general. It also promises perpetual reuse and no waste by connecting the work producing technosphere with nature's geosphere and biosphere, as well as proposing an ecological loop, as depicted in Figure 5. In short it is a recipe for long-lasting and complete sustainability, at least ecologically and materially speaking (McDonough and Braungart, 2002; and Cairns and Jacobs, 2017). Proponents draw on examples of the looseness of fit of downtown Manhattan buildings and particularly lofts with capacities of redundancy, convertibility, and adaptability as architectures of perpetual beginnings. Often this sentiment is also combined with 'design for deconstruction' in the literal sense of disassembly of structures, thus preserving materials and capacities for reuse, often despite difficulties associated with various jointing technologies (Keeler and Burke, 2009; and Bougdah and Sharples, 2010). Present in such proposals is a form of biocentric argument centered on the idea of some 'technological nutrient' much like nutrients in the chemical composition and metabolism of nature. Also in play are copious system boundaries defining consumption-waste loops to the exclusion of all waste losses. Perhaps not surprisingly there are critics of the very possibility of 'cradle-to-cradle' forms of artifactual production, one of whom is Frederich Schmidt-Bleek from the Wuppertal Institute for Climate, Environment and Energy, who perceives 'cradle-to-cradle' production as being entirely impractical and even falling into the biometric trap of viewing micro-reductions of objects into 'technical nutrients' capable of re-absorption much as equivalents in biological systems. Instead, Scmidt-Bleck advocates 'dematerialization' as a reduction in total material flows in any given system of design, production, and consumption and makes use of the metric of 'material intensity per source' (MIPS), much as the early summary measures presented here earlier (Schmidt-Bleck, 2008). As Cairns and Jacobs point out, advocates of 'cradle-to-cradle' accounts of production are also forgetting the necessary historical aspects of material flows. For instance, the quartz that became part of the cement that bound the bricks in the Manhattan structures as a 'technical nutrient' later became part of the 'gentrification nutrient' as they put it, with entirely different and somewhat untoward consequences (Cairns and Jacobs, 2017). In short, a vision of endless beginnings with 'cradle-to-cradle' concepts of production becomes confronted with a more likely collection of environmental stressors that at least incrementally can be materially and positively resolved from 'cradle-to-grave,' as work in this volume suggests.

However, the matter of close approximation to cradle-to-cradle formulations can also be considered, particularly in the prolonged reuse of stone buildings in compact city areas. For instance, given the simultaneous application of repair, refurbishment, reuse and demolition with material recycling and mining, how close can one come to zero material weight of material remaining unused, landfilled, or otherwise disposed of? To answer such a question, sensitivity testing across

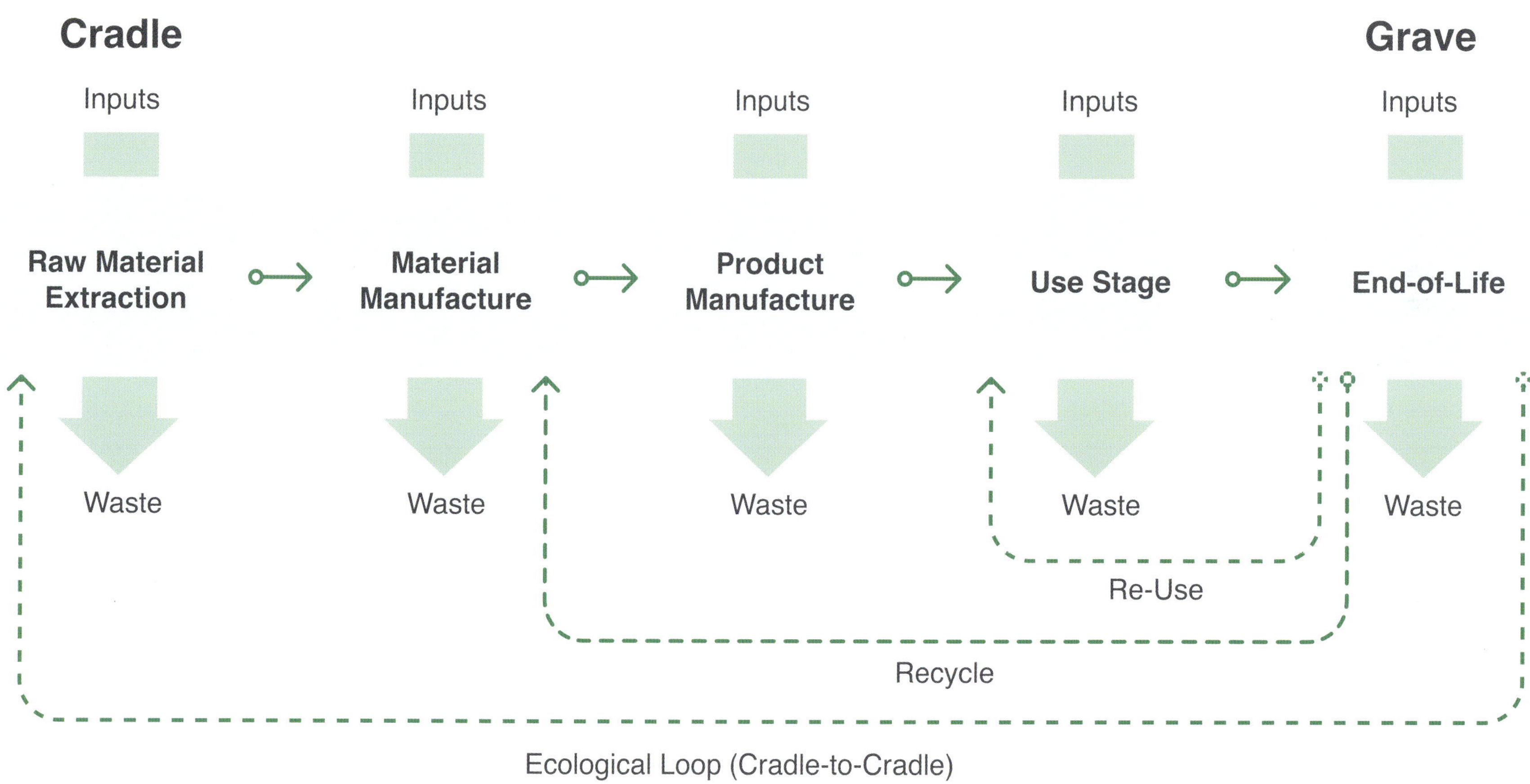

FIG. 5: Ecological Loop of Cradle-to-Cradle.

obsolescence rates of varying magnitude and proportions of recycling rates can be simulated and estimated within the stock-flow modeling methodology.

Returning to the formal expression of the mass conservation principle of the metabolic model described in Chapter 3, this approximation can be expressed as follows:

$$\sum_{p}^{1} M_{Input} = \sum_{q}^{1} M_{Output} + M_{Storage}$$

Where

$$M_{Storage} = \sum M_{Stock} + \sum M_{Waste}$$

Also

$$M_{Unused} = \sum_{p}^{1} M_{Imput} - \sum_{p}^{1} M_{Imput} \times \% \text{ Recycled} + \text{Other}$$

Where Other is unrecycled component of collected waste

$$M_{Unused} = \% \text{ of } \sum_{p}^{1} M_{Imput}$$

If

$M_{Unused} \leq 1\%$ of the cradle to cradle is approximated.

In this research, as elsewhere, rates of recycling and resource recovery vary both by location or context and by specific material. Generally, there are two forms of collected waste involved here: Municipal Solid Waste (MSW), or aspects of it, and Construction and Demolition Waste (C&D). Perforce of government regulation and industrial development, most rates have risen appreciably over the years, though they commonly remain modest under many current circumstances, especially in the face of perceived existential threats from global warming and resource loss, perceived by the likes of Paul Crutzen (Steffen, Crutzen and McNeill, 2007). Among the four contexts discussed here, Paris appears to lead the way in technological improvement, followed by Westwood in the U.S., and then Chencun, and La Campana

FIG.6: Obsolescence Rates and Material Recycling Rates for Paris.

in Monterrey, Mexico. For many basic building materials like concrete and steel, recycling and resource recovery rates of up to 95% appear to be demonstrable, at least in the relatively near-term future. Though again varying by location, material specific recovery improvements, like Smartcrusher BV - initially developed in The Netherlands - are a case in point for concrete. Similarly, asphalt in a reclaimed condition can be readily re-used as a foundation, fill and road material. Other materials, like steel, for example, benefit from deconstruction and source separation of up to 90 to 95%. Wood can also be highly recyclable and reuseable depending upon the context and local building technology, though this is less the case in places like Westwood in the U.S., where it is frequently combined too closely with insulation, paint and glue, as compared, say, to in Paris, where it is contaminated in its original

Paris, France

Obsolescence Rates	Current
Life cycle collective housing haussmanian buildings and equiva	200
Life cycle collective housing post-1920 buildings (y)	100
Life cycle commercial haussmanian buildings and equivalent (y	200
Life cycle commercial and public post-1920 buildings (y)	91
Life cycle industrial post-1920 building (y)	100
Life cycle agriculture pre-1920 building (y)	200
Life cycle agriculture post-1920 building (y)	100

Rates of Recycling	Current	Aimed	Best	Incorporated	Methods
% paper	32.0%	68.0%	91.0%	71.0%	Reuse and Recycling (post recovery)
% plastic	14.0%	68.0%	73.0%	14.0%	Reuse and Recycling (post recovery)
% aluminum	59.0%	70.0%	70.0%	50.0%	Reuse and Recycling (source separation)
% iron	95.0%	95.0%	95.0%	43.0%	Reuse and Recycling (source separation)
% steel	62.0%	85.0%	85.0%	43.0%	Reuse and Recycling (source separation)
% limestone	60.0%	90.0%	90.0%	8.0%	Deconstruction, Reuse and Recycling
% concrete	60.0%	90.0%	90.0%	8.0%	Reuse and Recycling (Smartcrushing)
% asphalt	60.0%	90.0%	95.0%	20.0%	Reuse and Recycling
% wood	61.0%	85.0%	98.0%	46.0%	Deconstruction, Reuse and Recycling
% glass	59.0%	68.0%	79.0%	63.0%	Reuse and Recycling (post recovery)
% clothing	20.0%	68.0%	73.0%	NA	Reuse and Recycling (source separation)
% rubber	55.0%	68.0%	73.0%	NA	Reuse and Recycling (source separation)

Westwood,MA,USA

Obsolescence Rates	Current
Life cycle suburbian buildings and equivalent (y)	61
Life cycle commercial and public buildings (y)	53.68
Life cycle industrial building (y)	0
Life cycle agriculture building (y)	0

Rates of Recycling	Current	Aimed	Best	Methods
% paper	68.2%	50.0%	75.0%	Deconstruction and Recycling
% plastic	8.0%	50.0%	75.0%	Deconstruction and Recycling
% aluminum	17.2%	50.0%	90.0%	Reuse and Recycling (source separation)
% iron	33.1%	50.0%	90.0%	Reuse and Recycling (source separation)
% steel	33.1%	50.0%	90.0%	Reuse and Recycling (source separation)
% limestone	30.0%	50.0%	50.0%	Deconstruction and using projected methods
% concrete	30.0%	50.0%	95.0%	Reuse and Recycling (Smartcrushing)
% asphalt	30.0%	50.0%	90.0%	Reuse and Recycling
% wood	17.1%	50.0%	50.0%	Deconstruction and Source Separation
% glass	25.0%	50.0%	90.0%	Reuse and Recycling (source separation)
% clothing	14.7%	50.0%	50.0%	Source Separation
% rubber	18.2%	50.0%	50.0%	Source Separation
% tar	30.0%	50.0%	50.0%	Source Separation
% sand	30.0%	50.0%	75.0%	Deconstruction and Recycling
% copper	30.0%	50.0%	90.0%	Reuse and Recycling (post recovery)
% gypsum	14.0%	50.0%	50.0%	Deconstruction and Recycling
% ceramic	12.0%	50.0%	50.0%	Deconstruction and Recycling
% brick	12.0%	50.0%	90.0%	Deconstruction and Recycling
% stone	30.0%	50.0%	95.0%	Deconstruction and Reuse
% minerals	30.0%	50.0%	50.0%	Deconstruction and Recycling
% gravel	30.0%	50.0%	75.0%	Reuse and Recycling (post recovery)

FIG.7: Obsolescence Rates and Material Recycling Rates for Westwood.

Chencun, China				
Obsolescence Rates	**Current**			
Life cycle suburbian buildings and equivalent (y)	35			
Life cycle commercial and public buildings (y)	35			
Life cycle industrial building (y)	35			
Life cycle agriculture building (y)	35			
Rates of Recycling	**Current**	**Aimed**	**Best**	**Methods**
% paper	20.00%	65.00%	65.00%	Reuse and Recycling (source separation)
% plastic	20.00%	65.00%	65.00%	Reuse and Recycling (recovery)
% aluminum	20.00%	65.00%	65.00%	Reuse and Recycling (recovery)
% iron	20.00%	65.00%	90.00%	Reuse and Recycling (source separation)
% steel	20.00%	65.00%	90.00%	Reuse and Recycling (source separation)
% limestone	20.00%	65.00%	65.00%	Deconstruction and Recycling
% concrete	20.00%	65.00%	95.00%	Reuse and Recycling (Smartcrushing)
% asphalt	20.00%	65.00%	90.00%	Reuse and Recycling
% wood	20.00%	65.00%	65.00%	Deconstruction and Recycling
% glass	20.00%	65.00%	90.00%	Reuse and Recycling (source separation)
% clothing	20.00%	65.00%	65.00%	Reuse and Recycling (recovery)
% rubber	20.00%	65.00%	65.00%	Reuse and Recycling (source separation)
% brick	20.00%	65.00%	90.00%	Deconstruction and Recycling
% clay pots	20.00%	65.00%	90.00%	Recycling

FIG.8: Obsolescence Rates and Material Recycling Rates for Chencun.

Monterrey, Mexico				
Obsolescence Rates	**Current**			
Life cycle suburbian buildings and equivalent (y)	20			
Life cycle commercial and public buildings (y)	20			
Life cycle industrial building (y)	20			
Life cycle agriculture building (y)	20			
Rates of Recycling	**Current**	**Aimed**	**Best**	**Methods**
% paper	52.00%	65.00%	65.00%	Reuse and Recycling(Recovery)
% plastic	57.00%	80.00%	80.00%	Reuse and Recycling(Recovery)
% aluminum	5.00%	35.00%	35.00%	Reuse and Recycling(Recovery)
% steel	5.00%	48.00%	90.00%	Reuse and Recycling (source separation)
% concrete	5.00%	48.00%	95.00%	Reuse and Recycling (Smartcrushing)
% asphalt	5.00%	48.00%	90.00%	Reuse and Recycling
% wood	5.00%	35.00%	35.00%	Reuse and Recycling(Recovery)
% glass	5.00%	40.00%	40.00%	Reuse and Recycling(Recovery)
% rubber	52.00%	52.00%	52.00%	Reuse and Recycling(Recovery)

FIG. 9: Obsolescence Rates and Material Recycling Rates for La Campana.

condition. Still other materials, like glass, for instance, remain highly recyclable but face major issues leading to lower recycling rates due to contamination and haphazard handling of material during demolition. Again, deconstruction and source separation strategies, such as those mandated by organizations like the Massachusetts Environmental Protection Agency, allow for up to 90% recovery.

From the perspective of embodied energy associated with the various material elements of each of the Sankey material flow analyses, results are presented in Figures 10, 11, and 12 and generally derived from reputable or acknowledged sources, at least for initial inputs (Te Herenga Waka-Victoria University of Melbourne, 1997). The overall totals for all four cases of settlements are shown here in Figure 12. In all cases except Chencun, recycling and resource recovery added around 40 percent to the weighted average totals. The case of Chencun, at around only a 26% markup, appears to be due to the preponderance of clay materials, including reusable pots, in the production process. At the more fine-grained level of specific building materials, initial embodied energy increases due to recycling and resource recovery are presented in Figure 11. For instance, the embodied energy of paper rises around 40% from 12.6 MJ/kg to 17.64 MJ/kg. For aluminum it rises by about 5% from 154 MJ/kg to 161.7 MJ/kg.

However, amid all these data and rising rates of obtainable recycling and resource recovery, close approximations to cradle-to-cradle formulations can also be considered, as was done in this research. As shown in Figure 12, assuming existing obsolescence rates, demonstrated upward movements in recycling and resource extraction rates were computed across total masses of collected

Contexts	Obsolescence Rates	Recycling Rates		
		Current	Aimed (2030)	Optimal
Westwood, MA, USA	61	25.20%	50.00%	77.32%
Paris, France	200	55.40%	84.19%	88.14%
Chencun, China	35	20.00%	65.00%	83.44%
Monterrey, Mexico	20	6.60%	46.00%	86.90%

FIG. 10: Current and Recycled Levels by Settlement Types.

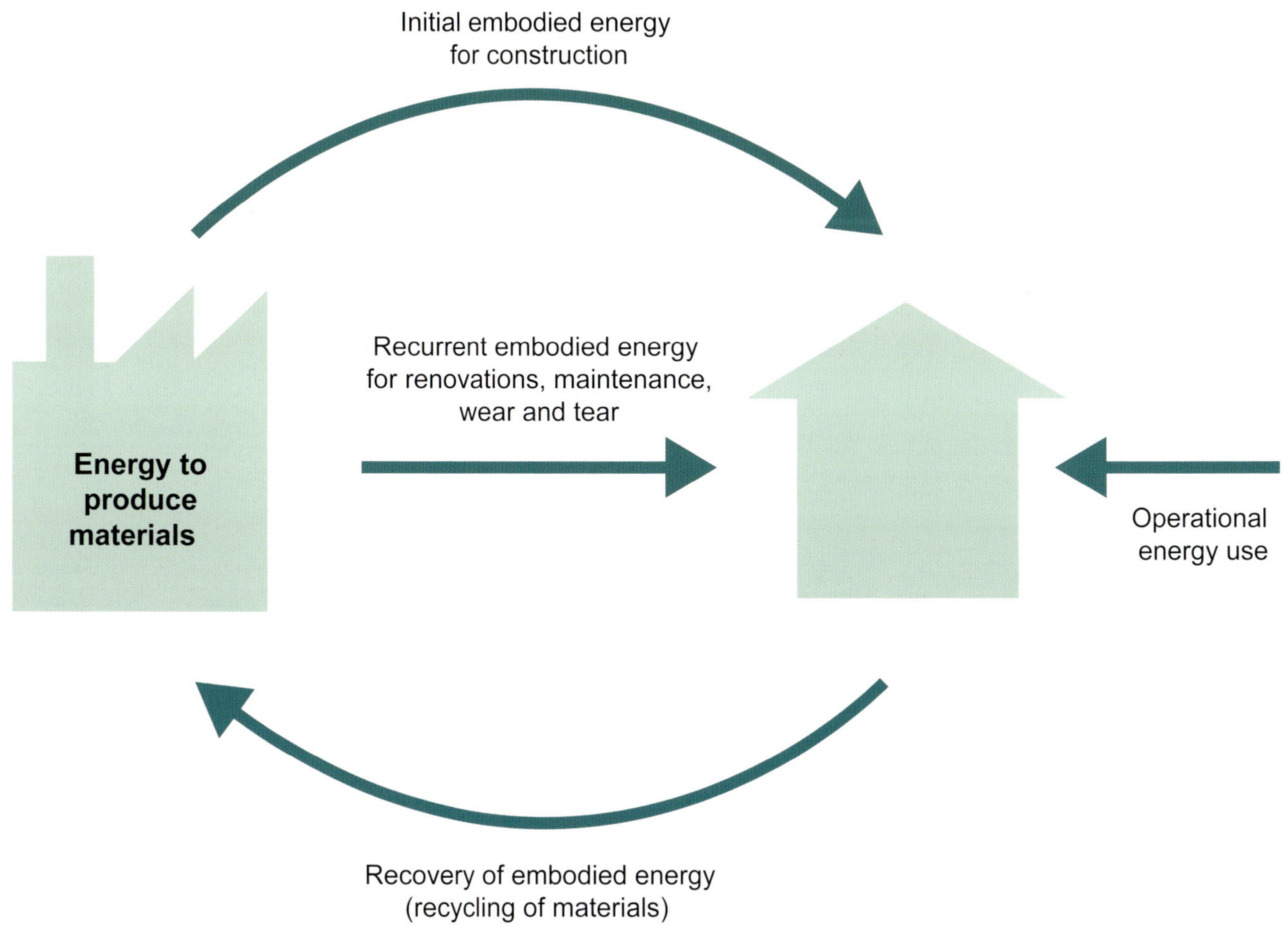

FIG.11: Embodied and Operational Energy in Buildings.

weight is much the same at around 280 metric tons per unit, though higher in embodied energy by a factor of an additional 80 percent or so. This, in turn, suggests that the predominantly stone construction is superior, also further suggesting that stone lattice buildings, even of high-rise dimensions, are probably a way to go. Second, urban design proposals, like those contemplated for the Champs-Élysées in Paris, among others, also do not seem likely to yield significant reductions in environmental impact, though providing improvements in other ways. Third, the major areas of significant gain in lessening embodied energy and carbon are to be found in operational aspects of urban development like mobility, adaptive reuse, heating and cooling, waste management, and so on. Throughout, reductions in the use of fossil fuels must be pursued vigorously, and up to a 50 percent or so improvement might be anticipated by shifts to electrical power, along with renewable sources. Fourth, as just noted, risk, resilience, adaptation and sustainability are inextricably linked with the obsolescence of buildings and constructed environments and the longevity that can be extracted from them, including reasonable dematerialization and resource recovery. For whatever reason, buildings must die in order for settlements to continue to flourish and adapt to new prevailing conditions. Also, as shown earlier, recycling rates must be improved.

07

Buildings in the Anthropocene

This final chapter draws together several different arguments and lines of thought running through the book. First are the interdependencies involved in metabolic concepts and the complexity of networks entrained in cities and buildings. Next is the growth of architecture alongside the science of building, as one of emergence and consideration of architecture within metabolic regimes.

1. A Different Green Revolution

We have become all too familiar with the haunting statistic: buildings produce nearly 40 percent of the world's yearly carbon emissions. Manifestos have been issued, government protocols passed, and problem-solving packages offered by profit-seeking corporations. Architects and builders appear to be keenly aware of the skeletons in their closet so, we assume, radical change must be in the works.

The trouble is that environmental activism in architecture cancels the founding myth of the discipline. Ancient temples to modern buildings value permanence. Schools, hospitals, and our precious homes are meant to stand against time. Writings on architecture from Philo of Alexandria through Vitruvius, Gaudi, and Le Corbusier boost the mythology of architecture as a discipline that transforms quotidian materials into something timeless. The architect is elevated into a magician of matter as she not only transforms it into something beautiful and utilitarian but stretches its temporal limits.

The lesson of our planetary collapse is that the limits of the material should be heeded. This book has documented methods to rethink how human settlements function as interconnected ecological systems. The metabolic analysis of cities demonstrates that materials systems are interdependent and thereby have limits that count.

Every part of a system is related to other parts: the birds drink the water, the water evaporates into clouds that carry moisture over buildings that in turn catch the run-off, channeling it down gutters, onto streets, and back to river deltas that eventually run into the vast oceans. This interconnected view is the metabolic regime.

Architecture is deeply embroiled in the metabolization of our planet. Every building relies on the transformation of minerals into manufactured goods like steel beams, glass curtain walls, cement floor boards, plastic conduit, enamel toilet bowls, door hardware, window latches, bathroom tiles, paint on the walls, and a plethora of embedded machines to keep all systems working. As such, contemporary buildings consist of countless environmental processes that react with their surroundings in ways that can be maintained and renovated or demolished and recycled.

This book is about how to view this new paradigm and what to derive from it. Sophisticated modeling of settlements and their built fabric has revealed numerous startling discoveries. Meticulous calculation of the material stock of each built environment followed by an analysis of how these stocks change or flow from one state to another informs a metabolic understanding of our world. This new regime offers a different, albeit complex, yet much more precise way of seeing the way in which our built environment acts as an interface for humans and their environment.

The deep interdependence of our environment reveals how the objects and structures that surround us are built on networks of production and extraction that span borders and millennia. These are human-initiated networks that cross over and entangle natural systems. Understanding the outcome of these entanglements is the aim of a metabolic flow analysis. Far from being isolated processes, these activities feed and complement each other, creating a tangled web of people, materials, and data. This book visualizes the cycles of production that form a built environment, crisscrossing time and space, to tell a story of buildings as nodes in changing an interdependent materialist milieu.

During the postwar years, just when these networks were getting more entangled and complex, architects of the modern movement fetishized the machine. There were a few design practitioners on the peripheries such as Cedric Price who paid attention to the form and extent of the systems being propagated by new technologies. The challenge was in being able to find what they were looking for. Computational power had not caught up with the intuition that complex global networks run through our cities and buildings. They knew that all things were connected but could not, as Einstein noted, discern in what way the connections were made, buoyed, and modified.

2. Can Complex Machines Float?

Architecture has always grown alongside science. Our machines have influenced and shaped how we build our world. The standard scientific method of taking a large problem and breaking it down into pieces to be studied, understood, and then recombined characterizes the way we have conceived of our built environment. Alberti's canonical theory of architectural beauty presents the architect as a virtuoso composer of parts to make a whole.

Just as physics discovered the quanta, biology the fluid dynamics of cell growth, astronomy dark matter, and even social sciences the complex behavioral patterns affecting events like markets and elections, architecture has come around to the understanding that buildings are not just machines for living but physical interfaces to manage ourselves in our environment. These physical interfaces despite seeming to be solid and static are anything but. They grow, change, die, and have afterlives.

The desire to find a different way to think about part-to-whole relationships runs through the study of metabolic flows our built environment. Instead of thinking of a building in terms of discrete individual elements such as the beams and structure that add up to make a building, metabolic analysis reveals that buildings are in fact not equal to the sum of their parts but greater.

Computational power has enabled the study of processes responsible for natural phenomena like waves on the ocean, light in the atmosphere, clouds on the horizon, and particles at sub-

FIG. 1: Lurin Eco-Industrial Park.

atomic scales. Scientists employing these tools discovered that each of these phenomena were the result of processes occurring over time and revealing different emergent properties. The parts of any object, system, or building were engaged in the process of making. The thing we call the whole is not exactly a fixed object but a state. Any object is, therefore, more a function of time than a spatial arrangement of its constitutive parts.

The implication of this shift in understanding has been to rewrite the notion of stability. Objects in a field, buildings in a landscape, and individuals in a community are not static. Their interaction generates new proprieties in the system that propels changes to both the part and the whole, thereby producing a dynamic scenario referred to as emergence.

The conundrum for any science of emergence is whether one can predict that which will emerge. All living systems reach balance. This tendency towards a state of stability is the homeostatic urge of every system. Homeostasis enables adaptation in constantly fluctuating environments.

Architecture is a practice of rigor, rules, and procedures meant to establish a balance between competing forces such as outside/inside, hot/cold, chaos/stability. The physicality of an architectural structure gains meaning and significance in how it performs as a homeostatic interface. We control and manage our environment through architecture which in turn means we control and manage ourselves through this medium.

The lesson is not to confuse the liquid for the container as Heidegger legendary once said; architecture's function is not to embody a preassigned functional program or to materialize an individual's vision. Architecture is bigger than us. Buildings are homeostatic interfaces because they serve as platforms for both interaction and characterization. In other words, things and people not only move through buildings but are understood by how they do so. The temporal processes that work on agents also affect buildings so that death, decay, and perhaps even recreation happen to our built environments just like they do to the organic material that flows through them.

This information-rich, process-driven model of the reality is what marks a metabolic worldview of architecture. Seen from this perspective our built environment starts to look like both a thing and a phenomenon. It is composed of physical elements whose in buildings is an attempt to come to terms, both quantitively and qualitatively, with the commotion caused by architecture on our planet.

Scientists in the middle of the 20th century started describing this new class of order as complexity. They used the term to describe systems that were lifelike, information-rich, and subject to phase transitions. Immense resources were devoted to the modeling of these processes. Research focused on the way parts in a system were affected by information from the environment, processed it, and in turn responded. Change in a system came to be understood as a response of the system to its surroundings.

Amidst this fluctuating field of flows and interdependencies, architecture insists on the value of scale. Buildings realize granular episodes of matter that put a frame around the smooth connectivity of ecological space. Rooms have bounding walls. Structures have rhythmic bays of beams and columns. Any work of architecture, from a home to a school, an office, airport, or museum is an aggregation of space at very precise scales with desired functions. The coordination of scale to manage flow might be the great technological innovation of architecture.

So, when the legendary physicist Max Planck said that "everything only looks like a thing from its own particular vantage point," he might have been thinking like an architect in the metabolic regime. The flows that constitute our world are active phenomena that disappear. The physical stuff that we assemble to hold our lives function because they compartmentalize scale in orders that manage and control the disappearing flows. The parts and the whole relationship are integral to how buildings can be made and yet the terms of the relationship need not be hierarchical. In the metabolic regime, relations have bidirectional interactions wherein all parts of the whole influence each other. Parts relate in regime open to the flows of metabolic processes.

Computer scientists and mathematicians have named this relationship-based type of order a 'holarchy.' In a holarchy, many things that we might normally think are impossible became probable. The idea that water might have memory becomes thinkable. The immunologist Jacques Benveniste proposed the idea in 1980. Our built environment connects us to the natural systems on our planet as well as the personal histories of communities.

Langston Hughes's poem, "The Negro Speaks of Rivers" begins: "I've known rivers: I've known rivers ancient as the world and older than the flow of human blood in human veins. / My soul has grown deep like rivers. / I bathed in the Euphrates when dawns were young. / I built my hut near the Congo and it lulled me to sleep. / I looked upon the Nile and raised the pyramids above it . . ." The poet can feel how his body is connected to other natural bodies, rivers both close and far, carrying stories of pasts to help us understand our own place.

This expansive empathy can be a way forward for architecture. It offers a way of thinking about our buildings that imagine many sources, many associations, joining our bodies to that of both the living and the dead, the organic and non-organic, so that we find beauty in the ways we relate to things.

08

Figures Within text

Chapter 5: Data and Proxies.

Chapter 6: Aspects of Improved Performance.

Chapter 7: Building in the Anthropocene.

09

References

Adiony, E. 2014. "Negros First Cyber Center Area Now a Commercial Zone." , May 28.

Ali, S. H. 2014. "Social and Environmental Impact of Rare Earth Industries." , vol. 3, no. 11, pp. 123–124.

Alves, B. 2024. "Recycling Rates of MunicipalmWaste in France, 2000-2023" *Statista*. March 28.

Appel, M.; A. Francis; A. Payne; A. Tanimoto and S. Mouw. 2024. *The State of Recyclin: The Present and Future of Residential Recycling in the US*. Washington, DC: The Recycling Partnership.

APUR, 2010. *Reference Documents*, Paris Atelier Parisian d'Urbanisme.

Appel, E., N. Karzis and M. Murphy, 2022. The Landscape for Commercial Property Conversion in New York City". N.Y. *Furman Center Policy Review*.

Aragon-Correa, J.A.; Matıas-Reche, F.; Senise-Barrio, M.E. 2004. Managerial discretion and corporate commitmentto the natural environment. *Journal of Business Research*, vol. 57, pp. 964–975.

Baccini, P. and P. H. Brunner. 2012 (1991). *Metabolism of the Anthroposphere: Analysis, Evaluation, Design*. Cambridge, Mass: The MIT Press.

Bahers, J. B; A. Athanassiadis, D. Perrotti, et al. 2022. "The place of space in urban metabolism research: Towards a spatial turn." *Landscape and Urban Planning*, vol. 221.

Barrows, D. P. 2022. *A History of the Philippines*. Charleston, S.C.: Legare Street Press.

Baron, D. P., Harjoto, M. A. and Jo, H, The Economics and Politics of Corporate Social Performance. 2009. Stanford University Graduate School of Business Research Paper. no. 1993, Rock Center for Corporate Governance at Stanford University Working Paper no. 45.

Benoit, B. ed. 2017. *Paris Haussmann: Modele de Ville – A Model's Relevance*. Zurich: Park Books.

Bertalanffy, L. von, 1968. *General Systems Theory: Foundations, Developments, Applications*. New York: George Braziller.

Billig, M. S. 2003. *Barons, Brokers and Buyers: The Institutions and Culture of Philippine Sugar*. Honolulu: University of Hawaii Press.

Brenner, N. 2019. *New Urban Spaces: Urban Theory and the Scale Question*. New York: Oxford University Press.

Bougdah, H. and S. Sharples, 2010. *Environment, Technology and Sustainability* London: Taylor and Francis.

Cairns, S. and J. M. Jacobs, 2017. *Buildings Must Die: A Perverse View of Architecture*. Cambridge, Mass: The MIT Press.

Candelaria, J. L. P. and V. C. Alporha. 2018. *Readings in Philippine History*. Manila: Rex Book Store.

Census of Popilation. 2020. "Region VI: Total Population by Province, City, Municipality and Baranguay." *Philippine Statistical Authority. July*.

Constantino, R. 2010. *A History of the Philippines*. New York: Monthly Interview Press.

Cossu, R. and I. D. Williams, 2015. "Urban Mining: Concepts, Terminology, Challenges." *Waste Management*, vol. 45, November, pp. 1–3.

Cushman-Roisin, B. and J-M Beckers. 2011. *Introduction to Geophysical Fluid Dynamics*. Cambridge, Mass: Academic Press.

Decoster, F.; D. Klouche and C. Poulin. 2021. *La ville, matière vivante L'AUC*. Paris: Parentheses.

Derrible, S.; L. Cheah.; M. Arora, and L. W. Teow,2021. "Urban Metabolism" in S. Geetmann, C. Petitand A. Sraffans, eds. *Urban Informatics*. Singapore; Springer, pp. 85-114.

Desmond, A. J. 1989. *The Politics of Evolution, Morphology, Medicine and Reform in Radical London*. Chicago: University of Chicago Press.

Diamantidis, D; M. Sykora and E. Bertacca. 2018. "Obsolescence Rate: Framework, Analysis and Influence on Risk Acceptance Criteria," in R. Caspeele, L. Tasrwe, and D. Frangopol, eds. *Life Cycle Analysis and Assessment in Civil Engineering: Towards an Integrated Vision*, London: CRC Press, pp.1–8.

Dick, S. J. 2008. *Remembering the Space Age*. Washington, DC: NASA Office of External Relations, History Division.

Diehl, A. F. and K. S. Smith, 2003. *Mining Waste Overview*. Washington, DC: U.S. Department of the Interior.

Doussard, C.; P. G. Rowe, P. Detavernier, and B. Guo, 2024. "Metabolism and Urban Design: Insights from the Champs-Élysées, Paris." *Sustainability*. vol.16, no.2, 603.

Dovey, K. 2012. "Informal Urbanism and Complex Adaptive Assemblage." *International Development Planning Review*, vol. 34, no. 4, pp. 349–368.

Earls, A. R. 2002. *Route 128 and the Birth of the Age of High Tech*, Boston: Arcadia Publishing.

Elliot, T.; J.B. Almenar, S. Niza, et al. 2019. "Pathways to Modeling Ecosystem Services Within an Urban Metabolic Framework." *Sustainability*. Vol. 11, no. 10, 2766.

Embudo, F. L. 2019. "Bacolod City: Top Model City is Most Livable Center in the Philippines." *Manila Times*, October 3.

Environmental Protection Agency, 2020. *Advancing Sustainable Materials Management: 2018 Fact Sheet: Assessing Trends in Materials Generation and Management in the United States*, Washington, DC: The EPA.

Food and Agriculture Organization. 2018. *World Food and Agriculture Statistical Pocket Book*. Rome: United Nations Food and Agriculture Organization.

Forman, R. T. T. 1995. *Land Mosaics: The Ecology of Landscapes and Regions*. Cambridge: Cambridge University Press.

Frosch, R. and N. E. Gallopoulos. 1989. "Strategies of Manufacturing." *Scientific American*, vol. 261, no. 3, pp. 144–152.

Gade, D. W. 1996. "Carl Troll on Nature and Culture in the Andes." *Erdkunde*, vol. 50, no. 4, pp. 301–316.

Gomez, C. 2023. "Turning the Tide in Waste: Mexico's Journey Towards a Circular Economy". *Circular Innovation Lab*, June 21.

Haeckel, E. 1866. *The General Morphology of Organisms*. Berlin: Reiner.

Hagen, J. B. 1992. *An Entangled Bank: The Origins of Ecosystem Biology*. New Brunswick: Rutgers University Press.

Hanlon, B.; J. R. Short and T. J. Vivino. 2010. *Cities and Suburbs: New Metropolitan Realities in the U.S.* New York: Routledge.

Harman, G. 2011. *The Quadruple Object*. U.K.: Zero Books.

Harman, G. 2018. *Object-Oriented Ontology: A New Theory of Everything*. London: Penguin.

Herment, A. 1856. *Les Fetes Publique*. Paris: Editions Aux Bureaux de L'artiste.

Hewitt, P. G.; J. A. Suuchcock and L. A. Hewitt, 2009. *Conceptual Physical Science Explorations*. Pittsburgh: Benjamin Cummings.

Huang, B.; X. Wang, H. Kua, et al. 2018. "Construction and Demolition Waste Management in China by the 3R Principle." *Resources, Conservation and Recycling*, vol. 129, pp. 36–44.

ISO, 2015. *General Principles on Reliability of Structures* (2394). Geneva: USO.

Jarrasse, D. and J. Lebar, 2007. *Grammaire des Jardins Parisienne: De Heritage de rois aux creations Contemporaine*. Paros: Parigramme.

Jiang, H.; Z. Sun, H. Guo, et al. 2021. "A Standardized Data Set of Built-Up Areas of China's Cities with Populations over 300.000 for the Period 1995-2015." *Big Earth Data*, vol. 6, no. 1, pp. 103–126.

Kangas, P. 2004. "The Role of Passive Electrical Analogs in H. T. Odum's System Thinking." *Ecological Modeling*, vol.178, no. 1, pp. 101–106.

Keeler, M. and B. Burke, 2009. *Fundamentals for Integrated Design for Sustainable Building*. Washington, DC: U.S. Green Building Council.

Kennedy, S. T; S. Pincetl and P. Bunje, 2011. "The study of urban metabolism and its application to urban planning and design." *Environmental Pollution*, vol. 159, no. 8-9, pp. 1,965–1,973.

Klouche, D. 2010. "News of Urban Utopias? The Grand Pari(s) of the Ile-de-France Metropolis." *Cities*, Vol. 2, no. 42, pp. 97–114.

Koolhaas, R. 1978. *Delirious New York*. New York: Oxford University Press.

Kneese, A. 1988. "The Economics of Natural Resources." *Population and Development Review*, vol, 14, pp. 281–309.

Johnson, R. D. and C. Holbrow. eds. 1977. *Space Settlements: A Design Study*. Washington, DC: NASA Scientific and Technical Information Office.

Lasafin, D. and A. Jovel. 2018. "When Was Bacolod City Born, Really?" *Panay News*, October 28.

Lavoisier, A. L. and P. S. Laplace. 1780. "Memoir on Heat." *Academie Royale des Sciences*, pp. 4–355.

Lavoisier, A. 1789. *Elements of Chemistry*. Chicago: Henry Regnry Company.

Lawson, R.; R. P. Hostetler and B. H. Edwards. 2008. *Essential Calculus Early Transcendental Functions*. New York: Grove Atlantic.

Lederer, J.; D. Laner and J. Fellner, et al. 2014. "A Framework for the Evaluation of Anthropogenic Resources Based on Natural Resource Evaluation Concepts." *Proceedings of the 2nd Symposium on Urban Mining*, Bergamo, Italy.

Liu, X.; Y. Song, et al. 2015. "Understanding Urban China with Open Data." *Cities*, vol. 47, pp. 3–29.

MacArthur, R. and E. I. Wilson. 1967. *The Theory of Island Biogeography*. Princeton: Princeton University Press.

Mandelbrot, B. and R. L. Hudson. 2008. *The (mis)Behavior of Markets: A Fractal View of Risk, Ruin and Reward*. London: Profile Books.

Marquise du Châtelet. 1744. *Dissertation sur la Nature et Propagation du Feu*. Paris: Chez Prault.

McDonough, W. and M. Braungart. 2002. *Cradle to Cradle: Remaking the Way*. New York: North Point Press.

McGee, T. G. 1991. *The Emergence of Desakota Regions on Asia: Expanding a Hypothesis*. Honolulu: University of Hawaii Press.

Morton, T. 2010. *The Ecological Thought*. Cambridge, Mass.: Harvard University Press.

Morton, T. 2013. *Hyperobjects: Philosophy and Ecology after the End of the World*. Minneapolis: University of Minnesota Press.

Neuenschwander, D. E. 2011. *Emmy Noether's Wonderful Theorem*. Baltimore: Johns Hopkins University Press.

Neponset River Watershed Association, 2017. *Natural History*. Canton, Mass.: Neponset Tiver Watershed Association.

Odum, E. P. and H. T. Odum. 1953. *Fundamentals of Ecology*. Philadelphia: Saunders Publishing.

Oswald, F.; P. Baccini and M. Michaeli. 2003. *Netstadt*. Berlin: Springer Science and Business Media.

Packard, V. 2007 (1957). *Hidden Persuasion*. New York: Ig Publishing.

Parsons, H. M. 1974. "What Happened at Hawthorn?" *Science*, vol. 183, pp. 922–932.

PCA-STREAM, 2020.*Champs-Élyséess Histoire et Perspectives*. Paris: Editions Pavillion de L'Arsenal – PCA Editions.

PCA-STREAM, 2023. *Etude Urbane Reenchanter les Champs-Élyséess*. Paris: Cahier 3.

Peterson, B. 2020. "Alluvial vs. Sankey Diagrams." *Datasmith.org*.

Peterson, K. S. 2022. "Fact Check: False Claims That Wind Turbine Generators Last Only 3 to 4 Years." *USA Today*, August 15.

Popov, G.; B. K. Lyon and B. Hollcroft, eds. 2016. *Risk assessment: A Practical Guide in Assessing Operational Risks*. New York: John Wiley and Sons.

Porter, Michael E., and Claas van der Linde. 1995. "Toward a New Conception of the Environment-Competitiveness Relationship." *Journal of Economic Perspectives*, vol.9, no.4, pp. 97–118.

Pound, R. and F. E. Clements. 1998. "A Method of Determining the Abundance of Secondary Species." *Minnesota Botanical Studies*, vol. 2, pp. 19–24.

Rackwitz, R. 2000. "Optimization." *Structural Safety*, vol. 22, no. 2, pp. 27–60.

Real, L. A. and J. H. Brown, eds. 2012. *Foundations of Ecology: Classic Papers and Commentaries*. Chicago: University of Chicago Press.

Rendgen, S. 2018. *The Minard System: The Complete Structural Graphics of Charles Joseph Minard for the Ecole Nationale des Ponts et Chausses*. New York: Princeton Architectural Press.

Richardson, C. and M. J. Pisani. 2012. *The Informal and Underground Economy of the South Texas Border*. Austin: University of Texas Press.

Rickerts, R. E. 2005. *The Economy of Nature*. New York: W. H. Freeman.

Rowe, P, G. 1991. *Making a Middle Landscape*. Cambridge, Mass.: The MIT Press.

Rowe, P. G. 2017. "Causes of Obsolescence and Building Futures" in H. Kara; L. A. Villoria and G. Georgoulous, eds. *Architecture and Waste: A (Re) Planned Obsolescence*. Barcelona: Actar.

Rowe, P. G. and L. Hee, 2019. *A City in Blue and Green: The Singapore Story*. Singapore: Springer Nature.

Saich, T. 2001. *Governance and Politics in China*. New York: Palgrave.

Saldaña, C. G. 2000. *Financial Mergers in the Philippine Setting: Text and Cases*. Quezon City: AFA Publications.

Santorio, S. 1737. *Medicina Statica: Being the Aphorisms of Sanctorius*. London: W. and J. Newton.

Schmidt, M. 2008. "The Sankey Diagram in Energy and Material Flow Management." *Journal of Industrial Ecology*, vol. 12, no. 2, pp. 173–188.

Schnid-Bleck, F. 2008. "Factor 10: The Future of Stuff." *Sustainability Science Practice and Policy*, vol 4, no. 1, pp. 1–4.

Shmurak, S. 2018. "Paris is Building the Eco-Community of the Future Right Now: Here's How." *Inspiration*, 25 January.

Sijakovic, M. and A. Peric. 2014. "Recycling Architecture: The Redefinition of Recycling Principles in the Context of Sustainable Architectural Design." *Proceedings REAL CORP 2014*, pp. 467–476.

Sollis-Sarether, H. and P. Gottschalk. 2010. "The Modeling Process of Stage Models." *Journal of Organizational Computing and Electronic Commerce*, vol. 20, no. 3, July.

Sommerville, E. 2023. "Ignore the Doom-mongers, Says New UN Climate Change Chief." *The Telegraph*, 31 August.

Steffen, W.; P. J. Crutzen, and J. R. McNeill, 2007. "The Anthropocene: Are Humans Now Overwhelming the Great Forces of Nature." *Ambio*, vol. 30, no. 8, pp. 614–621.

Steinitz, C. 2012. *A Framework for Geodesign: Changing Geography by Design*. Redlands: Esri Press.

Stern, J. 2019. *Global Outsourcing, Local Transformations: Transnational Business and Urban Restructuring in the Philippines*. PhD Dissertation, Cambridge, Mass: Harvard Graduate School of Design.

Storper, M. 1887. *The Regional World: Territorial Development in a Global Economy*. New York: The Guilford Press.

SURBA. 2023. *Introduction: Selected Projects*. Website at surba.net.

Thomsen, A. and K. van der Flier. 2011. "Obsolescence and the End of Life of Buildings." *Proceedings, Management and Innovation for Sustainable Built Environments*, Amsterdam, June 20–23.

Trevino, R. 2020. "This Initiative Seeks to Support the Area Where 'Ya No Estoy Aqui' was Filmed." *Connect*, September 28.

Turvey, S. T. and J. J. Crees. 2019. "Extinction in the Anthropocene." *Current Biology*, vol. 29, no. 19, pp. 982–986.

United Nations, 2015, *Sendai Framework for Disaster Risk Reduction (2015–2030)*. New York: United Nations General Assembly.

United Nations Convention on Climate Change. 2023. "Views on the Elements for the Consideration of Output Components of the First Global Stock Take." *2023 Synthesis Report on GST Elements*. Geneva: UN Convention on Climate Change.

Walker, E. 2002. "How Long Do Solar Panels Last?" *Enerysage*, December 12.

Waxman, O. B. 2019. "Meet 'Mr. Earth Day' The Man Who Helped Organize the Annual Observance." *Time*, April 19.

Wente, van A. 1990. "Scientific and Technical Information Management." *Government Information Quarterly*, vol. 7, no. 2, pp. 149–167.

Westwood, MA. 2023. *On-line Assessment Database*. Boston: Vision Government Solutions.

Wolman, A. 1965. "Metabolism of Cities." *Scientific American*, vol. 203, no. 3, pp. 179–190.

World Commission on Environment and Development, 1987. *Our Common Cause.* Oxford: Oxford University Press.

Wu, B. S. and D. Sui, 2015. "Modeling Impact of Globalization on Desakota Regions: A Case Study of Taipei Metropolitan Area." *Environmental Planning B: Planning and Design*, pp. 1–21.

Wuebbles, D. J.; D. W. Fahey, K. A. Hibbard, et al. 2017. *Climate Science Special Report, vol. 1*, Washington, DC: U.S. Global Change Research Program.

Xiqo, S., H. Dong, Y. Geng and M. Broader. 2018. "An Overview of China's Recyclable Waste Recycling and Recommendations for Integrated Solutions". *Resources, Conservation and Recycling*, vol. 134, pp. 112-120.

Zapata, J. 2019. "Pocket Parks in La Campana in Monterrey, Mexico." *Critical International Planning*, pp. 1–8.

Zhang, Y.; Z. Yang, and X. Yu. 2015. "Urban Metabolism: A Review of Current Knowledge and Directions for Future Study." *Environmental Science and Technology*, vol. 49, no. 19, pp. 11,247–11,263.

10
Illustration Credits

Chapter 1.

Fig. 1 Drawn by authors based on NOAA Climate.gov graph, 2024.

Fig. 2 Drawn by authors based on NOAA Climate.gov graph, 2024.

Chapter 2.

Fig. 1 Public Domain.

Fig. 2 Public Domain.

Fig. 3 Drawn by authors.

Chapter 3.

Fig. 1 Drawn by authors based on McGee, 1991.

Fig. 2 Drawn by authors.

Fig. 3 Drawn by authors.

Fig. 4 Drawn by authors based on Sankey's original diagram, 1898.

Fig. 5 Drawn by authors based on Laurence Livermore National Laboratory, 2022.

Fig. 6 Drawn by authors.

Fig. 7 Drawn by authors.

Fig. 8 Drawn by authors.

Chapter 4.

Fig. 1 Drawn by authors.

Fig. 2 Drawn by authors.

Fig. 3 Copyright and courtesy of Philippe Guignard, Air Images.

Fig. 4 Drawn by authors based on Paris Métropole Aménagement.

Fig. 5 Creative Commons, 2016.

Fig. 6 Drawn by authors based on Polymago and Paris Batignolles Aménagement, 2015.

Fig. 7 Drawn by authors using SankeyMATIC.

Fig. 8 Drawn by authors using SankeyMATIC.

Fig. 9 Drawn by authors using SankeyMATIC.
Fig. 10 Drawn by authors.
Fig. 11 Drawn by authors.
Fig. 12 Copyright and courtesy of Nancy Fyler, Neponset River Watershed Association.
Fig. 13 Copyright and courtesy of Nancy Fyler, Neponset River Watershed Association.
Fig. 14 Drawn by authors using Google Earth Map, 2023.
Fig. 15 Copyright and courtesy of Zillow.com, 2023.
Fig. 16 Drawn by authors using SankeyMATIC.
Fig. 17 Drawn by authors using SankeyMATIC.
Fig. 18 Drawn by authors using SankeyMATIC.
Fig. 19 Drawn by authors using SankeyMATIC.
Fig. 20 Drawn by authors using SankeyMATIC.
Fig. 21 Drawn by authors.
Fig. 22 Drawn by authors using Google Earth Map, 2023.
Fig. 23 Drawn by authors using Google Earth Map, 2023.
Fig. 24 Copyright and courtesy of Peter G. Rowe.
Fig. 25 Copyright and courtesy of Peter G. Rowe.
Fig. 26 Copyright and courtesy of Panorama Global, 2020.
Fig. 27 Copyright and courtesy of Peter G. Rowe.
Fig. 28 Drawn by authors using SankeyMATIC.
Fig. 29 Drawn by authors using SankeyMATIC.
Fig. 30 Drawn by authors using SankeyMATIC.
Fig. 31 Drawn by authors using SankeyMATIC.
Fig. 32 Drawn by authors using SankeyMATIC.
Fig. 33 Drawn by authors.
Fig. 34 Copyright and courtesy of McGee, 1991.
Fig. 35 Drawn by authors using Google Map, 2023.
Fig. 36 Drawn by authors.
Fig. 37 Copyright and courtesy of Boya Guo, 2023.
Fig. 38 Copyright and courtesy of Boya Guo, 2023.
Fig. 39 Copyright and courtesy of Boya Guo, 2023.
Fig. 40 Copyright and courtesy of UN Food and Agricultural Organization, 2016.
Fig. 41 Drawn by authors using SankeyMATIC.
Fig. 42 Drawn by authors using SankeyMATIC.
Fig. 43 Drawn by authors using SankeyMATIC.
Fig. 44 Drawn by authors using SankeyMATIC.
Fig. 45 Drawn by authors using SankeyMATIC.
Fig. 47 Drawn by authors.

Fig. 48 Drawn by authors.
Fig. 49 Drawn by authors.
Fig. 50 Drawn by authors.
Fig. 51 Drawn by authors.
Fig. 52 Drawn by authors.
Fig. 53 Drawn by authors.
Fig. 54 Drawn by authors.
Fig. 55 Drawn by authors.
Fig. 56 Copyright and courtesy of Peter G. Rowe.
Fig. 57 Copyright and courtesy of Peter G. Rowe.
Fig. 58 Copyright and courtesy of PCA-STREAM, 2023.
Fig. 59 Copyright and courtesy of PCA-STREAM, 2023.
Fig. 60 Copyright and courtesy of PCA-STREAM, 2023.
Fig. 61 Copyright and courtesy of PCA-STREAM, 2023.
Fig. 62 Drawn by authors using SankeyMATIC.
Fig. 63 Drawn by authors using SankeyMATIC.
Fig. 64 Drawn by authors using SankeyMATIC.
Fig. 65 Drawn by authors using Google Map, 2023.
Fig. 66 Drawn by authors based on Westwood, Massachusetts Tax Assessment, 2023.
Fig. 67 Copyright and courtesy of Zillow.com, 2023.
Fig. 68 Copyright and courtesy of Zillow.com, 2023.
Fig. 69 Drawn by authors using SankeyMATIC.
Fig. 70 Drawn by authors.
Fig. 71 Copyright and courtesy of Pariscorporatehousing.com, 2023.
Fig. 72 Drawn by authors using Google Eath Map, 2023.
Fig. 73 Drawn by authors using Google Map, 2023.
Fig. 74 Drawn by authors using SankeyMATIC.
Fig. 75 Copyright and courtesy of CAZA.
Fig. 76 Copyright and courtesy of CAZA.
Fig. 77 Drawn by authors.
Fig. 78 Copyright and courtesy of CAZA.
Fig. 79 Copyright and courtesy of CAZA.
Fig. 80 Copyright and courtesy of CAZA.
Fig. 81 Copyright and courtesy of CAZA.

Chapter 5.

Fig. 1 Drawn by authors based on Westwood, Massachusetts Tax Assessment, 2023.
Fig. 2 Drawn by authors.

Chapter 6.

Fig. 1 Drawn by authors.
Fig. 2 Drawn by authors based on Risk Assessment Graph, Wikipedia, 2006.
Fig. 3 Drawn by authors based on Thomsen and van der Flier, 2011.
Fig. 4 Drawn by authors based on Thomsen and van der Flier, 2011.
Fig. 5 Drawn by authors based on Paintsforlife.eu, 2021.
Fig. 6 Drawn by authors.
Fig. 7 Drawn by authors.
Fig. 8 Drawn by authors.
Fig. 9 Drawn by authors.
Fig. 10 Drawn by authors.
Fig. 11 Drawn by authors based on Murray Hall.
Fig. 12 Drawn by authors.

Chapter 7.

Fig. 1 Copyright and courtesy of CAZA.
Fig. 2 Copyright and courtesy of CAZA.

11

Acknowledgments

This book started some ten or more years ago, as Claire Doussard would spend time with Peter Rowe during the summer at Harvard University's Graduate School of Design, of which she was a recent graduate, working on what became her doctoral dissertation at La Sorbonne in Paris, and resulting in the stock-flow metabolic models that undergird this text. Later on, as this work progressed to its present state it also permeated several institutions on both sides of the Atlantic. At Harvard University it received several faculty grants from the Center for Green Buildings and Cities and the stalwart support of its leader Professor Ali Malkawi and able assistances by Robert Marino and Samantha Sarafin. Also, at Harvard we are indebted to the likes of Hanif Kara, Rahul Mehrotra and Michael Manfredi for their opinions and insights. At the École Spéciale d'Architecture in Paris, its Dean, Marie-Hélène Contal not only shared our interest in sustainable design but also graciously assisted us in our efforts. In addition, Etienne Riot, among others, at PCA-Stream, a prominent Parisian urban design firm helped us with practical problem-solving approaches. Also, at Pratt Institute in New York, several students in its graduate program engaged with us in their research projects and one, dealing with modular commercial infrastructure authored by Tamar Salinger, appears in this text. At SURBA, where Carlos Arnaiz and Peter Rowe practice, a number of staff members participated in the design of projects using various ideas of this metabolic research. These included Gaby San Roman in the early days, Nick Schlegel especially with regard to the use in a church and botanical garden complex made of volcanic ash bricks, Katie Cantine, particularly with regard to the new Government Center in Victorias, and Gerardo Corona with regard to the Negros Island projects in The Philippines. Finally, several of our SURBA clients warrant special acknowledgement for their steadfast support of novel adaptive reuse and recycling approaches to projects. These certainly include the brothers Benetiz, Albee the Mayor of Bacolod and Kiko the Congressman of the 3rd District in Negros Island, as well as the priests of the De La Salle congregation.

12

Author Profiles

Peter G. Rowe is the Founding Chairman of SURBA – The Studio for Urban Analyses in Brooklyn New York. He is also the Raymond Garbe Professor of Architecture and Urban Design at Harvard University's Graduate School of Design. as well as a Harvard University Distinguished Service Professor. He served for twelve years as Dean of the Faculty of Design at Harvard and prior to that as Chairman of the Urban Planning and Design Department and Director of the Urban Design Department. Before joining the Harvard faculty, Rowe was the Director of the School of Architecture at Rice University and Vice-President and Research Director of the Rice Center and Environmental Program Director of the Southwest Center for Urban Research. He is also the holder of several honorary professorships in East Asia and was Education Director of the Aga Kahn Center for Culture as well as the author, co-author and editor of numerous articles and some thirty-one books. Including recently Design Thinking band Story Telling in Architecture, Architectural Territories in East Asia and When Urban Design Comes to Ground, also with Carlos Arnaiz. He received his Bachelor of Architecture degree from Melbourne University and his Mater of Architecture urban Design from Rice University.

Claire Doussard is a landscape engineer, urban planner and urban researcher specializing in sustainable urban planning. environmental and technological urban innovations and research development strategies. She has an active consulting practice and is the author and co-author of several books and numerous scientific articles, as well as communications for the general public. Doussard also holds an appointment as Assistant Professor at the École Spéciale d'Architecture from 2020 to present, as well as serving as a researcher at the University of Lausanne from 2022-2023 and a lecturer at the University of Cergy-Pontoise from 2017-2020. She was also President of Scago SAS in Paris from 2015-2018. Doussard earned a PhD in Planning from the University of Paris, Sorbonne in 2017, a Master's degree in Urban Planning and Design at Harvard University's Graduate School of Design in 2014 and a degree in the Management and Landscape Engineering from Agrocampus Ouest in 2012.

Carlos Arnaiz is an architect, educator, writer and urban design consultant. He is the CEO and Co-Founder of SURBA – the Studio for Urban Analyses in Brooklyn New York and the Founder and Principal of CAZA a Brooklyn-based architectural firm. He is also an Adjunct Assistant Professor at the Graduate School of Architecture and Urban Design at Pratt Institute where he has taught courses on the history of ideas about the city, modular construction technologies and design studios on a variety of topics including urban housing. His professional experience ranges from the design and construction of large-scale mixed-use complexes to strategic sustainability plans for new urban developments in various parts of the world, in addition to numerous single buildings for public and private sector clients. Arnaiz is the recipient of numerous design awards and holds a Master of Architecture degree from Harvard University's Graduate School of Design with honors and a Bachelor of Arts degree, magna cum laude, from Williams College.

Index

Note: Page numbers in italics refer to figures